The Urban Book Series

The Urban Book Series is a resource for urban studies and geography research worldwide. It provides a unique and innovative resource for the latest developments in the field, nurturing a comprehensive and encompassing publication venue for urban studies, urban geography, planning and regional development.

The series publishes peer-reviewed volumes related to urbanization, sustainability, urban environments, sustainable urbanism, governance, globalization, urban and sustainable development, spatial and area studies, urban management, transport systems, urban infrastructure, urban dynamics, green cities and urban landscapes. It also invites research which documents urbanization processes and urban dynamics on a national, regional and local level, welcoming case studies, as well as comparative and applied research.

The series will appeal to urbanists, geographers, planners, engineers, architects, policy makers, and to all of those interested in a wide-ranging overview of contemporary urban studies and innovations in the field. It accepts monographs, edited volumes and textbooks.

Indexed by Scopus.

Meike Levin-Keitel · Lukas Behrend

The Topology of Planning Theories

A Systematization of Planning Knowledge

Springer

Meike Levin-Keitel
Department of Spatial Planning
TU Dortmund University
Dortmund, Germany

Lukas Behrend
Department of Spatial Planning
TU Dortmund University
Dortmund, Germany

ISSN 2365-757X ISSN 2365-7588 (electronic)
The Urban Book Series
ISBN 978-3-031-37856-0 ISBN 978-3-031-37857-7 (eBook)
https://doi.org/10.1007/978-3-031-37857-7

This Springer imprint is published by the registered company Springer Nature Switzerland AG
The registered company address is: Gewerbestrasse 11, 6330 Cham, Switzerland

Paper in this product is recyclable.

Contents

Analytical Index

1. Just a Word in the Beginning...

Why is this book necessary? Not only are planning theories undervalued in planning sciences, they are also an essentially anarchic endeavour: Criticism of existing approaches and efforts to establish a foundation are more prevalent than constructive contributions to further development.

2. Planning Practice and Practices of Spatial Planning

There is no single, clearly definable spatial planning practice when it comes to planning theories. As a collective term for a large variety of practices and expertise, it is important to make them explicit as a necessary prerequisite for communication about planning in general and a theoretical reflection on planning approaches in particular.

2.1 Spatial Planning—a Dazzling Area Calling for Action

Planning is a multifaceted field of action and encompasses very different practices. The actions and the knowledge required for doing planning are correspondingly diverse. Who plans what and how does this relate to a narrow and broad understanding of planning?

2.2 Planning Practice—a Look at Practices of Spatial Planning

Planning practice, whether defined as the epistemic object of planning research or as the constituent subject of planning, seems insufficiently specified. The understanding of planning practices depends on the approach, the chosen theory.

2.3. Diversity of Understandings of Planning—and Fundamental Differences

Planning practice is a broad subject and it entails some implicit basic assumptions. This requires an examination of the effects of theory and practice as an inseparable unit.

3. Planning Knowledge

The epistemological question of knowledge of spatial planning can be expressed in knowledge of the present and the future. Planning theories reflect the ambivalence of the essential future orientation of planning knowledge and the importance of the normativity of this knowledge.

3.1 Knowledge About Knowledge in General

Knowledge is traditionally defined as true and reasoned belief - a widely criticised circumstance. However, the fact that something is said to be true does not mean that it actually is or remains true. Various forms of knowledge acquisition contain basic epistemological concepts.

3.2 Knowledge in Spatial Planning

Planning knowledge describes the fundamental establishing of planning approaches. A distinction can be made between descriptive, analytical, predictive and normative knowledge, however, decisive seems to be its normative knowledge content.

4. Value Orientation in Spatial Planning

The field of action of spatial planning as an interface between future-oriented solutions to complex existing problems is the arena of ethical and moral orientations and alignments. However, planners cannot make these ethical decisions alone. Orientation towards a common good in particular requires further differentiation in this sense.

4.1 The Transition from Knowledge to Action

There is no direct link from an is to an ought, from existing knowledge to an distinct action. The question of what to do is therefore a completely different one than the question what we know. Accordingly, in between lies a grand field of values, including moral positions.

4.2 Moral Positions and Action-Orientation

The three positions of normative ethics for evaluating an action also differ significantly when it comes to planning: Virtue ethics, which primarily values intention, deontology, which addresses the action itself, and utilitarianism, which deals with its consequences.

4.3 Between Planning Knowledge and Democratic Action

There is no such thing as value-free knowledge. For planners, a conflict arises due to their technical knowledge or their specialist expertise and the political decision makers, who do not necessarily have to use this knowledge as a basis, since political decisions are mostly value-based judgements.

5. Planning Theories and Existing Systematisation Approaches

Theories reflect different bodies of knowledge and are based on characteristic assumptions and basic values. Previous systematisation approaches place significant importance on the historical genesis and changes in planning understandings as well as in the distinction between substantive and procedural theories.

5.1 Theories as Systematisation of Knowledge—Also in Planning?

Planning theories can be understood as condensed knowledge, as basic constructs of planning, but is this also the case in planning sciences? Approaches can be recognized, but a systematisation of the specific knowledge of planning is still required.

5.2 Systematisation of Planning Theories—and Their Relation to Planning Knowledge

Existing systematisations provide an overview of the different theoretical approaches over time or in practical research work. This does not yet correspond to a systematisation of planning knowledge with regard to scientific-theoretical foundations.

6. Topology of Planning Theories

The Topology of Planning Theories as a newly introduced systematisation builds on the characteristic features of planning knowledge: the location in the scientific field and the degree of action-orientation of planning theories. The topology outlines the diversity and importance of planning knowledge for the profession and discipline.

6.1 Basics of a Topology—From Genesis to Validity

The Topology as a systematisation of theoretical approaches of planning places individual approaches in relation to each other in the theoretical field. The characteristic location in theoretical fields and the degree of action-orientation in these theories form the basis of the Topology.

6.2 The Field of Planning Knowledge—an Epistemological Systematisation

The field of planning knowledge contains epistemological foundations, different theories each find their specific place. Basic classification is based on the question of how reality(ies) are grasped and what knowledge is contained in the theory.

6.3 The Degree of Action-Orientation in Planning Theories—Macro, Meso and Micro Level

The degree of hands-on learning as an omnipresent discussion in planning is considered a further determinant: from the social macro level to the (neo) institutional meso level, to an action-oriented micro level, all references to action are included.

6.4 The Application of the Topology—Examples, Potentials and Limits

The benefits and potentials of the Topology evolve in its application: Relating theories to one other (on the basis of their epistemological position) in their foundation (on the basis of action-orientation and their theories). With all its potentials and limits.

7. What can Planning Achieve?

The conclusions that can be drawn for planning as a discipline and professional field from a reorganisation of planning knowledge give rise to hope. The presentation of the diversity of planning knowledge beyond procedural theories and their genesis can serve as a basis for urgently needed spatial and planning expertise for the challenges of the 21st century.

Just a Word in the Beginning 1

Abstract

Why is this book necessary? Not only are planning theories undervalued in planning sciences, they are also an essentially anarchic endeavour: Criticism of existing approaches and efforts to establish a foundation are more prevalent than constructive contributions for enhancement.

Keywords

Spatial planning • Theoretical foundations • Planning theories

"I am daily ever more convinced that theoretical work accomplishes more in the world than practical work. Once the realm of representation [Vorstellung] is revolutionized, actuality [Wirklichkeit] will not hold out." Hegel in a letter to Niethammer 1808 (Butler and Seiler 1984, p. 179).

Theories as the starting point of any consideration We have been dealing with planning theories for several years now, and just as Hegel emphasises the imaginary power of theoretical work, we have also found in numerous discussions and reflections that it is precisely the theoretical perspectives that make up our reality and perception that determines our problems, their approaches to solving them and how challenges are dealt with in society. In doing so, theories serve as a lens, as glasses, through which we grasp the world out there, making it tangible and discussable, pouring it into constructs of realities and defining all our assumptions. One of the most profound experiences in dealing with planning theories has been the realisation that these theoretical lenses can be changed and the reach of these social constructions can be realised, through which the perception of the imaginable changes massively (once the realm of representation [Vorstellung] is revolutionized) and what we call reality appears completely different—*actuality [Wirklichkeit] will not hold out.* What can change through this shift in perspective is all-encompassing: the definition of the problem, the approach to the problem,

M. Levin-Keitel and L. Behrend, *The Topology of Planning Theories*, The Urban Book Series, https://doi.org/10.1007/978-3-031-37857-7_1

the selected methods and forms of knowledge, the possible solutions sought and to be achieved, and last but not least the basic understanding of what we call spatial planning. It is all the more astonishing that in many discussions and publications that deal with planning theories, it is degraded as being not so profitable in practice and rather a necessary evil for planning research. On the other hand, there are numerous difficulties and challenges that become apparent when one deals more closely with planning theories, primarily the disagreement in the scientific community about what spatial planning is and what planning theories should encompass. Friedmann summed this up in his much-cited article *Planning Theory Revisited* (1998), speaking of spatial planning as a contested form of what we want to theorise ("nature of the beast we wanted to theorise") and far beyond that, by unmasking the rejection of concrete definitions of planning and planning theories as perceived restrictions on freedom by planning scientists. The image he used, that planning scientists ride different horses in different directions towards the sunset and therefore everyone stays on their own saddle and no common perspective can be identified, also speaks for the controversial identification of what planning is. In this book, we would like to contribute to the epistemological foundation of what we call spatial planning. To be consistent with Friedmann's narrative, set a common anchor point for the discussions, the place of a campfire, where you can at least sit together and listen to other perspectives, maybe even discover other perspectives by getting on another horse. Because a change of perspective is always associated with a change of a theoretical approach, a change in perspective of reality is not an easy undertaking.

Theoretical work in planning From a large number of discussions and talks with scientists who deal with spatial planning in theory and practice, in universities and local authorities, in Germany and internationally, with benevolence and a critical distance, it became clear that it is time to reconsider planning and its scientific reflections again. Not because there is consensus and agreement on this, but rather because it seems that the fundamental basics of planning and thus the epistemological background of planning theories is hardly discussed and has not been entirely worked out. In other words, one could say that as planning scientists we are *speechless* when it comes to distinguishing our disciplinary planning knowledge from other forms of knowledge, positioning it as scientific knowledge and articulating it scientifically despite interdisciplinarity and close relation to action. But, what is our language? What does each individual mean when the word spatial planning is used? Which terms are implicitly manifested within it? And, what does this mean for the planning sciences and the science of planning?

Why this theoretical foundation is (not) necessary In discussions about what planning theories are and what they are supposed to achieve, many voices have their say that deny the relevance of a more in-depth consideration of planning knowledge. In our opinion, many statements that we have heard and taken note of already reflect large parts of the problem: Why bother with such questions? Why determine the understanding of planning? Why deal with questions like planning as a scientific discipline? With this book, we would like to contribute to reflecting on planning knowledge in terms of scientific theory and thus to systematically discuss the basics of onto-epistemological approaches. The Topology of Planning Theories as a systematisation deals with ontological, epistemological and ethical perspectives of planning knowledge and underlines a special focus on its action-orientation. What do we need that for? In recent years, we have heard many arguments as to why we do not need to deal with the fundamentals of scientific theory. Perhaps some of them sound familiar to you. We would first like to present you with a list of three counter-arguments why you should ***not*** read this book. These are statements by planners from practice and planning scientists. If you are inclined to agree with these arguments, then you are right and can safely put this book aside. If you have even the slightest doubt or curiosity about how further study can develop our knowledge of spatial planning, we invite you to reflect with us on the knowledge and science of planning.

An initial statement as to why a continued study of the issue is not necessary is as follows:

> "I don't feel comfortable articulating an explicit understanding of planning that underlies my perspective; such a definition always excludes other understandings of planning as it then is very deterministic."

This a relatively frequently used argument as to why any concrete definitions of spatial planning and how planning can be understood are avoided. In doing so, it is neither necessary to establish ***one*** understanding of planning, i.e. to determine what planning is and what it is not, nor to establish this for all planners for all of time. Rather, it is of great importance to make explicit what understanding of planning a person has, not in a deterministic manner, but in a way that is comprehensible for the discussion partners, and thus it becomes clear which conclusions and insights can be drawn from this. Does spatial planning mean governmental action by a city planning office or the development of an urban garden by a civil society initiative? Do only formal planning instruments count as spatial planning tools or can these also be financial incentives to save space? All of these mostly implicit understandings and differences are neither wrong nor right, but are an essential basis for all further insights that result—both in research and for people in practice. This makes it clear that transparency about the respective understanding of planning *not* only does not have a deterministic effect, but also prevents further deterministic false conclusions, since the relation to the planning theories can be differentiated. Reflections on spatial planning that puts state-led action in the foreground differ significantly in problem definition, instruments and approaches from planning that also includes civil society initiatives. A clarification of which understanding a person follows therefore seems to be the basis of any discussion rather than a deterministic definition.

Another statement targeting the substantive level of planning theories and planning knowledge that we have heard in discussions is:

> "We do have theories on how a city or a region should be developed: Distance calculations, accessibility, equipment of infrastructures; all of these are the rules of the game, so to speak, according to which cities and regions are developed."

Even if there are a lot of rules of the game, especially in the operationalisation of planning, according to which cities and regions are developed, the further question arises as to what is behind these specifications, which theoretical assumptions are behind these calculations and on which societal, social scientific and basic economic assumptions they are based on. Not only theories and models on the distribution of infrastructures or the accessibility of individual locations are important, but also theories on how decisions on spatial development should be made, who is legitimate to do this and who should be involved and how. As mentioned in the argument, the models and legal requirements play an important role at the direct level of action. However, the knowledge on which these models are based is much more extensive and is based on different social ideas, technical innovations and possibilities as well as economic ideas. Since these can change constantly, it is important to be aware of these basics in order not to rigidly insist on distance calculations or infrastructure equipment, but to adapt and enhance them if necessary.

A very special feature of planning knowledge is the directly corresponding field of action of spatial planning. This direct action-orientation has a massive impact on planning theories:

> "For me, it is enough that spatial planning has a specific relation to action, which means that planning sciences already have a basis of legitimacy. Whether it's a discipline or not doesn't really matter anymore."

The question of whether spatial planning can be regarded as a scientific discipline or not has puzzled planning scientists for a long time. The underlying question is a question of the legitimacy of the planning sciences on the one hand (if it is not a discipline, then the legitimacy of the planning sciences is questioned), on the other hand a question of the inner structuring and systematisation of the knowledge of these planning sciences (what is the specific planning knowledge then and how is it different from other disciplines?, see Behrend and Levin-Keitel 2020). The argumentation presented here, that the legitimation basis for a scientific discussion of spatial planning can be derived solely from the specific relation to action, is misleading on several levels. When dealing with the question of whether the planning sciences form a discipline or not, it is neither an academic exercise nor a theoretical question of dedicated scientists. Being (or not being) a discipline has consequences, from recognition as a science to the question of how specific technical knowledge can be characterized and located in the scientific system. A negation of the question of representing a scientific planning discipline has so far largely shied away from the question of which knowledge, which ontological and epistemological foundations planning knowledge is based on—with the argument that this plays no crucial role. In addition, the specific action-orientation alone does not justify being a science, but just underlines the professional field itself. Here another overarching phenomenon in the planning sciences becomes clear, a strong action-orientation. Countless articles discuss practice, what theories seem unable to depict, or the gap between theory and practice and what that means for research and implementation of spatial planning. To get straight to the point, even if this seems to be the tenor in the planning sciences, we do not agree with this point of view of a gap between theory and practice. The relation to action and the relationship between theories and practices will keep us busy throughout this book and are considered an inherent part of the Topology of Planning Theories.

Cornerstones of the discussion and a topology The three counter-arguments—all taken from discussions with planning theorists—describe very impressively the area of conflict in which this book is thematically located: between (1) a more precise specification of planning knowledge, (2) the explicit basic assumptions and understanding of planning in planning theories and (3) their characteristic relation to action.

1. Scientific knowledge is presented in the form of theories, in our case planning theories and their description of a theoretical field. However, planning theories are not only selected partially and as examples in planning sciences, they are also a fundamentally anarchic undertaking: Criticism of existing approaches and a lack of a basis is made more often than constructive contributions are made to further development. Theories, no matter how specific or general, reflect different bodies of knowledge and are based on certain assumptions and core values (Barker and Kitcher 2014). These are rarely made explicit and discussed in planning sciences (Alexander 2016, p. 94). Previous structuring approaches place a lot of emphasis on the chronological classification and the historical change in planning understandings as well as on the distinction between substantive and procedural theories (Allmendinger 2017). However, this neither does justice to the knowledge inherent in planning theories (Davoudi 2015) nor does it help to answer important questions and connections, such as the ontological relationship between theories and reality/practice or the question of the ethical foundations of the planning discipline.
2. We see an urgent need to make the basic assumptions and planning understandings of theoretical approaches explicit and thus to enhance them—what kind of knowledge do we have on planning and on which theoretical assumptions is it based? In previous approaches to the systematisation of planning theories, dichotomies are usually opened up, such as in the comparison of substantive and procedural theories. As fruitful as a simplification and delimitation can be on a theoretical level, it reflects the connections in practice insufficiently and does not do justice to the interdependencies in theory. A first step in another direction can be an inventory in which the existing planning knowledge, which is organized and validated in planning theories, is recorded and systematised.

Accordingly, we are attempting a new approach to the systematisation of planning theories that takes into account the demands of a simplifying theory and the diverse interdependencies in practice—the Topology of Planning Theories.

3. The characteristic action-orientation of planning sciences and partly of planning theories is one of the key dimensions that has crucial meaning and is subject to specific interpretations, especially in the discourses. As already briefly discussed, there are extremely interesting and decisive interpretations and reflections in the action-orientation and the strong action-relation of planning sciences as well as planning theories: what constitutes this planning practice (Forester 2012); whether planning can be described as practice of knowing (Davoudi 2015); that the further development of theory is often driven to come closer to practice (Innes 1995); or specifically the gap between theory and practice is discussed and how large it may be (Alexander 1997). Just as prominently as these questions currently determine the discourse, we will also address the planning practice in this book in detail; in particular, its relationship to theory and science and in turn their ontological perspectives on the practice and reality of planning. However, we adopt a different basic attitude towards the relationship between theory and practice, a basic attitude that assumes a unity between these terms. If you describe theories as a lens or glasses with which you approach the reality out there, then theory and practice are two sides of the same coin, irrevocably connected and dependent on each other; one does not exist without the other. Even if you are not necessarily always aware of the theoretical basic assumptions, they still influence what you perceive as reality. This means that theories and practices are a direct part of an overall view and the question of the action-orientation of planning theories is completely different: Which theory has which relation to action and what is this (implicitly) based on? In how far can the knowledge contained in these theories then be classified? As a logical consequence, action-orientation is also an important component of our Topology of Planning Theories.

The Topology of Planning Theories The Topology of Planning Theories represents a reorganisation of knowledge for planning in the twenty-first century. Numerous challenges, be it the effects of climate change and adaptation and protection strategies or social polarisation and the question of appropriate decision-making, require to rethink and develop a reorientation of planning approaches. This book does not offer answers and solutions in the general sense, but asks questions, enables other perspectives on previous planning theories and their knowledge content and restructures them. Planning knowledge as well as the strong action-orientation of planning sciences are discussed, questioned and other perspectives on known approaches are articulated. Knowledge and action are closely interwoven, or as Friedmann put it, "We cannot wish not to know, and we cannot escape the need to act" (Friedmann 1987, p. 84). We present a *Topology of Planning Theories*, a systematisation of theoretical approaches that is neither right nor wrong, but does justice to the scientific claim of a disciplinary foundation and integrates the accumulated knowledge from the field of spatial planning. The distinction between procedural or substantial, project-related or strategic, moderating or sovereign, dichotomous pairs of opposites, should be avoided, as this propositions correspond to ideal–typical basic assumptions and can hardly be found in this analytical separation. On the contrary, the topology approach emphasises how certain elements relate to one another, not to be confused with a typology, which is about distinguishing and differentiating individual elements from one another.

Exchange between German and International discourse It is crucial to know that the first edition of this book has been published in German, strongly rooted in German literature, German-speaking discourses in planning practice as well as sciences, and thus linking the upcoming German planning theory discourse with the English-speaking community. We decided to publish our contribution in English as well to strengthen the exchange between separate and thus limited discourses of planning theories. Since the last few years we observe a growing interest in discussing and reflecting on planning theories, not only in the English-speaking community. We feel honoured to take up these plural views and break through the language barrier to enrich both the English-speaking discourses as well the German-speaking discourses on planning theories. So, in some chapters we refer a bit more to German literature than usually in the English-speaking academic realm.

The narrative of this book With this book, we are pursuing several goals that cannot be explained solely by presenting the Topology of Planning Theories. The basic idea is and was to deal more deeply with the onto-epistemological foundations of these theories. We structured the chapters according to the most important questions that are relevant for planning practice and at the same time for planning theories: What constructions of reality do we start from in spatial planning and what do we call planning practice? To what extent does this influence planning knowledge and what is considered specific about these forms of knowledge? What role do normative-ethical aspects play in our theories and how does this relate to planning-specific future and action-orientation? And, how do we define planning theories in the area of conflict between action-orientation and the classification of onto-epistemological perspectives? Due to its linear structure in its chapters, this book follows a rough classification of the aspects in practice, science and theory, but as already pointed out, we perceive these categories as only slightly differentiated in colloquial language and reject this form of simplification. We always use cross-references and take up certain arguments repetedly in the individual chapters in order to make clear the inherent and complex connection between practice, science and theories. The way in which practice is defined significantly influences the knowledge of spatial planning and the scientific reflection of or with these practices, they are significantly interdependent and condition each other. In detail, this means:

The consideration of what we call spatial planning traditionally begins with the question of being, the field of action, the much-cited planning practice (Chap. 2). This includes different inherent ontological perspectives of planning theories and the question of what this reality of planning practice is (object or subject of knowledge), how it is defined and what approaches there are to grasping planning as a practical action. Our contribution of an ontological perspective on planning serves to differentiate this planning practice and to reveal different understandings of planning—not to define them, but to give possibilities for a differentiated view and to get away from planning practice without specifying it in more detail.

If the object/ subject of knowledge is described, then in Chap. 3 the question of the possibility of knowledge in general and thus the question of the epistemological dimension of planning knowledge arises. It is advisable to take a look at the forms of knowledge and basic assumptions of planning-related knowledge and to classify them epistemologically and to pursue the question of what knowledge the science starts from. How is knowledge defined in spatial planning and what forms of knowledge are represented in planning theories? The strong focus on future-oriented knowledge of spatial planning leads us to the question of which future should then be the desired one.

The ethical-normative perspective on spatial planning in Chap. 4 is clearly based on its future orientation, it is a matter of looking at values on which planning actions and institutional contexts are based. How should planners act? What values are the planning processes based on? Accordingly, planning primarily deals with questions of normative ethics and addresses the practical relation of planning to action as well as the values that precondition the knowledge base that is believed to be secure. We propose a differentiation of ethical considerations, which distinguish and partly question the normative claims and interpretations, for example the orientation towards the common good.

In Chap. 5, we present theories in general and planning theories in particular as agglomerated knowledge of a discipline and clarify their fundamental importance as a store of knowledge. We then broaden the perspective and consider existing systematisation approaches of planning theories. How do these approaches structure planning theories and what are the consequences? The division of planning theories into substantial and procedural approaches, which is a well-established divide in planning sciences, also seems to be part of the current inadequacies and limited scope of planning theories.

In Chap. 6, after the presentation of common systematisation approaches in the previous chapter, we draft a new systematisation approach of planning theories—the Topology of Planning Theories. The Topology of Planning Theories is based on two distinctive characteristics of planning knowledge, the location in the field of planning knowledge and the degree of action-orientation of the planning theories—both characteristics that were identified in the previous chapters as essential and characteristic of planning knowledge and planning theories. In this way, the Topology of Planning Theories offers an approach that can depict planning theories both in terms of their breadth of content and their various levels of action.

Finally, in Chap. 7, we wonder what spatial planning with its inherent knowledge can aim for in the future. What do the descriptions mean for spatial planning as a discipline and professional field and what conclusions can be drawn from them with regard to the further development of planning knowledge organisation?

After all, spatial and planning expertise is urgently needed today, especially in the face of the great challenge of social transformation processes.

References

Alexander ER (1997) A mile or a millimetre? measuring the 'planning theory—practice gap. Environ Plan B Plan Des 24:3–6. https://doi.org/10.1068/b240003

Alexander ER (2016) There is no planning—only planning practices: notes for spatial planning theories. Plan Theory 15(1):91–103. https://doi.org/10.1177/1473095215594617

Allmendinger P (2017) Planning theory. Palgrave, London

Barker G, Kitcher P (2014) Philosophy of science—a new introduction. Oxford University Press, New York

Behrend L, Levin-Keitel M (2020) Planning as scientific discipline? Digging deep toward the bottom line of the debate. Plan Theory 19(3):306–323. https://doi.org/10.1177/1473095219897283

Butler C, Seiler C (1984) Hegel: The Letters, translation. Indiana University Press, Bloomington

Davoudi S (2015) Planning as practice of knowing. Plann Theory 14(3):316–331. https://doi.org/10.1177/1473095215575919

Forester J (2012) Learning to improve practice: lessons from practice stories and practitioners' own discourse analyses (or why only the loons show up). Plann Theory Practice 13(1):11–26

Friedmann J (1987) Planning in the public domain: From knowledge to action. Princeton University Press, Princeton/New Jersey

Friedmann J (1998) Planning Theory Revisited. Eur Plann Stud 6(3):245–253. https://doi.org/10.1080/09654319808720459

Innes JE (1995) Planning theory's emerging paradigm: communicative action and interactive practice—closing the theory-practice-gap. J Plann Educ Res 14(3):183–189

Planning Practice and Practices of Spatial Planning

2

Abstract

There is no single, clearly definable spatial planning practice when it comes to planning theories. As a collective term for a large variety of practices and expertise, it is important to make them explicit as a necessary prerequisite for communication about planning in general and a theoretical reflection on planning approaches in particular. Planning is a multifaceted field of action and encompasses very different practices. The actions and the knowledge required for doing planning are correspondingly diverse. Who plans what and how does this relate to a narrow and broad understanding of planning?

Keywords

Spatial planning • Planning practice • Narrow and broad understanding of planning

Different practices Spatial planning, more specifically urban and regional planning, includes a diverse field of action. Economic, ecological, sociodemographic or transformative processes in space, the preparation and development of plans and strategies for future developments or listening to other, sometimes contradictory arguments and their democratic negotiation processes, all this and more is addressed in urban and regional planning. Spatial planning seems to be happening everywhere, at all times, and within a variety of actors engaged in what we call spatial development. It is obvious that planning does not only take place in administrative offices and that it is not necessarily the sole task of the state to decide what urban or regional development can look like. A look at the history of planning outlines that the once strong role of planning as a public task, carried out by the planning administrations of the city, the region, the federal states, has changed massively: from a strong state with influential planning claims (*Godfather model of planning*) towards the state as one actor among others, with certain legally binding instruments, but dependent on social and political acceptance and support

M. Levin-Keitel and L. Behrend, *The Topology of Planning Theories*, The Urban Book Series, https://doi.org/10.1007/978-3-031-37857-7_2

(*multi-actor governance*). If a person speaks of planning practice as a field of spatial planning, it becomes obvious that this in turn consists of many different practices, institutions, strategies, actions and subject areas. It is worth taking a closer look at these different aspects and how they shape spatial planning in general.

2.1 Spatial Planning—A Dazzling Area Calling for Action

Definitions of spatial planning The field of activity of spatial planning is extremely diverse, planning practice is defined in the literature in a correspondingly abstract way: Planning as a "*linkage between knowledge and organized action*" (Friedmann and Hudson 1974, p. 2), i.e. the connecting element between knowledge and action, as an attempt to control the future ("planning as future control", Wildavsky 1973, p. 128) or "planning activity as practice of knowing" (Davoudi 2015, p. 317). Wildavsky presents a perspective on planning that describes it as future control: "planning is the ability to control the future by current acts." (Wildavsky 1973, p. 128) Already in this example, the universally valid description of something future and present becomes clear, no matter how these two aspects are linked. Brooks (2019, p. 9) understands planning as the process with which we try to shape the future, Gunder (2010, p. 299) describes planning as an ideology of how we define and use space. What these and other definitions from other fundamental works have in common (such as Friedmann 1987; Throgmorton 1992; Hartmann 2012) is a strong future orientation (it is not just about describing the present), which is accompanied by a direct normative action-orientation (for an overview, see Wiechmann 2008, p. 63 ff.). As already pointed out by Alexander (2016, p. 92), the problem with all these definitions is not that they are not true, but rather that they are too abstract to define planning practice. These definitions are hardly sufficient in delimiting what is meant (or not covered) by the term spatial planning in the specific statement. It is undisputed that the future orientation of spatial planning undermines a specific analytical understanding, "the object of planning, future action, routinely involves the unique and novel" (Forester 1982, p. 3). Nevertheless, or perhaps because of this, it is essential to explain the underlying understanding of planning. Otherwise, the term spatial planning runs the risk of becoming an empty signifier, a term devoid of content, the meaning of which is eroding and is lost in the multitude of implicit understandings.

Who is meant by spatial planning?

Planners in a narrow understanding of planning A significant difference between a narrow and broad understanding of planning lies in the question of who actually plans and who is included in the term planner. This is undisputed in the narrow understanding of spatial planning as a public task; here planners are people who perform sovereign tasks of spatial planning and order in the public service, for instance urban planners or regional planners in local authorities. However, also in other parts of the administration of regional authorities decisions are made about spatially relevant goals, effects and measures, through sectoral planning, social infrastructure or fiscal distribution. In the narrow understanding of planning, this planning approach is referred to as sectoral planning, as task-related planning by various administration units. The term sectoral planning is considered a vague collective term under which very different types of planning are summarised (water management planning, traffic planning, forestry planning). They are spatially relevant or spatially significant if they take up space or influence the spatial development or function of an area. In a narrow understanding, other actors and groups of actors are also involved in planning processes, through participation formats, cooperation or contractual cooperation. Politicians have a significant influence on planning by passing resolutions and setting directions. Strictly speaking, spatial planning in the narrower sense only represents the preparation of a decision, while the decision itself is based on political negotiations and is only one part of spatial planning in a broad sense.

Planners in a broad understanding of planning In a broad understanding of spatial planning, planning takes place everywhere, in public administrations, by civil society initiatives and citizens, by different economic actors. Spatial planning, understood here as spatially relevant actions that influence future design and use, includes all actions that are future-oriented, integrated and action-oriented. In a specific planning case or project, it is important to identify and analyse the respective actor constellations, since only certain actors from the spheres of state, economy and society are involved in certain planning processes (see Fig. 2.1). Spatial planning in a broad sense leads to a complex understanding of which actors are active in spatial terms: from the finance department of a city or the political social committee (state sphere) to protest organizations or the media and their reporting (society sphere) to private planning offices or enterprises (economy sphere). The challenge of identifying spatial planning and its actor constellations and grasping them in their complexity significantly determines the theoretical reflections and discourses.

Implications for planning theories In the German-speaking area, the pair of terms spatial planning and development has been introduced to differentiate between narrow and broad planning understandings (Selle 2014, p. 75 ff.). Although this defines the different spheres that represent plurality a priori, the minimum role of public administration to distinguish planning in a narrow or broad sense remains unclear. The pair of terms, spatial planning and development, was not able to establish itself internationally. With regard to theoretical approaches, it can be stated that planning theories either take the narrow understanding of spatial planning as a basis or explicitly take an expanded form of the governance perspective and differentiate it. The former include discussions about the role of planning or planners, questions of controlling spatial development and planning in the multi-level state system. Many theoretical approaches in planning sciences are based on this narrow understanding of spatial planning, considering planning as a public task of public actors. However, the underlying understandings of planning are rarely made explicit. The latter, generally referred to as governance approaches for the sake of simplification, mean established theoretical approaches that assume several

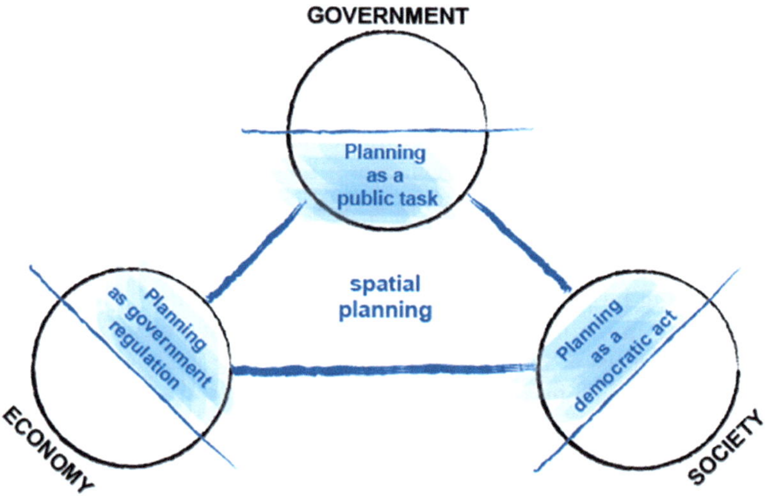

Fig. 2.1 Locally specific actor constellations from the three spheres of state, economy and society. (Own illustration)

planners, the government as a state actor, but also actors from society and business (McCann 2016). What is meant by governance here are different approaches that introduce a different point of view, an extension of government. This also includes, for instance, collaborative and communicative planning approaches or postmodern planning theories. Society and science are not only to be seen as being affected by planning, but also as an inherent part of planning—citizens and civil society initiatives plan, sometimes out of direct concern, sometimes as advocates of environmental concerns, as a sign of political positioning and democratic participation. Theoretical starting points for planning sciences are theories of democracy and social theories, which are transferred to spatial planning. Economic representatives often integrate their concerns through political representatives, the theoretical starting point is the relationship between planning and politics or power issues in planning. Intermediaries such as planning offices play an extra-ordinary role that has so far hardly been discussed in theoretical planning debates. These intermediaries provide technical support through expert opinions, sectoral or thematic concepts or process organization, but still follow an inherent economic logic (as a business) and have far-reaching influence in the design of processes.

What is planned?

Planning in a narrow understanding of planning Spatial planning in a narrow sense claims for a spatially integrated overall planning, i.e. planning that coordinates different interests for a specific place or region in an integrated manner. In a narrow sense, planning translates this claim by specifying physical space, specifying uses, prohibitions, and commands that are linked to this area and spatial units (especially in regional planning and local planning). The regulations not only relate to physical space, but also to social space, i.e. society and its individuals and their space-related actions as well as social objectives and principles. The starting point is a localisation and spatial consideration of different issues, which is expressed in physical-spatial regulations, but already defines a considerable influence on social actions and social objectives in the definition of the possible uses.

Planning in a broader understanding of planning While spatial planning in a narrow understanding is strongly linked to the question of the ability of the public sector to control spatial processes, planning in a broader understanding is more modestly oriented towards participation in a jointly designed spatial development. The focus of the question of what is planned shifts strongly to processes in social space, i.e. society, social groups, individuals and their negotiation processes. The variety of effects of spatial planning is expressed less in specific physical spaces, but rather in planning as joint learning processes in networks, as political discourse or as democratic goal setting in a specific planning area such as a city.

Implications for planning theories The question of what is planned is interpreted very differently in planning theory approaches. Here, as well, narrow and broad understandings of planning, the control and regulation of public space and its uses as well as the influencing of spatially relevant actions in social space, in society, are manifested. The strong reference to physical space, uses and functions is decisive for substantial planning theories such as the central place concept (Openshaw and Veneris 2003). While a narrow understanding of planning aims at theories with a dominant coordination claim of the public sector, the expansion of the understanding of planning results in a multitude of content plurality and the question of cooperation and communication among the actors involved. In addition, there is the question of integrating these plans into spatial, political, and thematic contexts; theoretical perspectives that are each introduced into the theoretical discourse of planning sciences from other disciplines. Depending on the understanding of planning, what is planned thus takes up the fundamental distinction between planning theories and substantive and procedural theories (Faludi 1973).

How is it planned?

Planning instruments in a narrow understanding of planning The question of how planning is carried out refers to their general validity beyond individual particular interests, their binding nature in social negotiation processes and their relevance in form of government, for instance a social market economy (in the United Kingdom, the Republic of Poland or the Federal Republic of Germany). In a narrow understanding of spatial planning, formal planning plays an important role, which is carried out by the state, based on laws and legal requirements, and has general validity. This form of formal planning is undisputedly at the heart of narrow spatial planning (at least in the German-speaking context), it fundamentally requires state representatives as actors who direct the planning process and set the plans, programs and measures through predetermined procedures. The interfaces to informal planning approaches are diverse, closely interwoven and difficult to define clearly in detail. In this way, e.g. informal nature conservation reports on a regional level can be used as a tool of integration of other aspects, in the light of species protection concerns of a subordinate planning level. Alternatively, urban development concepts serve as political guidelines that also affect development plans and settlement expansions. The strict dichotomy of formal and informal planning instruments hardly does justice to the diverse planning practice; both informal and formal planning instruments come into play in a narrow understanding of spatial planning.

Planning instruments in a broad understanding of planning In a broad understanding of spatial planning, the question of how planning is done can only be illustrated be exemplified by illustration without claim to completeness. In addition to formal instruments, this includes a large number of informal instruments and measures (Koch 2015). Informal planning approaches can take on different tasks: As socially legitimised knowledge, informal planning can depict target images, as reports and expertise in a larger spatial context, informal planning can be a knowledge base and important for the preparation of decisions, as sectoral concepts enable technical expertise and strategy formation. In addition, informal instruments can also be incentives (financial or otherwise) to promote certain behaviour. Due to the high variety of actors and their scope for action, this includes a large number of different instruments and approaches that have to be identified depending on the actors and goals of spatial planning.

Implications for planning theories The way in which planners plan differs significantly, depending on whether planning is understood as state-led action in cooperation with other actors or whether it is generally about spatially relevant action by all actors. In theoretical discussions, spatial planning is very closely linked to planning as a public task, with the focus on state action (e.g. *Planning in the Public Domain,* Friedmann 1987). This does not mean that only the administration plays a role as an actor, civil society initiatives can also be involved in formal planning processes, as representatives of public interests or in public participation formats. The same applies to economic actors who are bound by the specifications of legally binding planning, or planning offices who work on aspects of formal planning in greater depth as commissioned work by means of expert opinions or sectoral concepts. Nevertheless, state planning remains the initiator and responsible for formal planning instruments and measures and is ultimately the decision-maker. In planning sciences and planning theories, spatial planning as a state-led task represents the dominant perspective of planning theories, especially up to the postmodern era.

2.2 Planning Practice—A Look at Practices of Spatial Planning

Planning practice as a research object What is understood by spatial planning varies enormously depending on the understanding of planning and perspective on practice, the attempt at a generally valid definition inevitably poses great challenges for planning scientists. Practice is generally understood as a "set of activities linked by common understandings and abilities" (Alexander 2016, p. 92), or as epistemic knowledge-centred practice (ibid.). However, if, as in the use of the term planning practice, the epistemological object of planning is unclear or cannot be adequately identified, it also remains unclear what knowledge the practices draw on. The insufficient definition of the concept of planning consequently reveals the shortcoming of its practice: Does this mean the practice of planning as a public task or the practices of any spatially relevant action? The difference is significant depending on who performs, what is planned and how (Dobrucká 2016). In the literature, planning practice is defined as a set of actions, skills, a learning process, or institutionalised meanings, very general definitions without defining what the planning part is supposed to be. Alexander (2016) concludes that there is no such thing as planning per se, just a variety of spatial planning practices with clearly identifiable objects such as urban planning, traffic planning, etc. The city or traffic becomes an epistemological object, but not planning - strictly speaking, one can then speak of urban research or traffic research, less of planning research.

Planning practices as the subject of planning Planning scientists take a different perspective, considering planning practices as a constitutive element of the concept of planning. Planning, then, is what constitutes planning practices. Planning practices consist of a large number of activities, ranging from the creation of plans and the contribution with technical expertise to coordination within the administration to citizen participation processes and moderation. Planning can thus be described by formulating specific sub-areas, e.g. planning instruments, the planning object and the (institutional) context of planning ("tools, objects, context" Alexander 2016, p. 95). It remains unclear which understanding of planning is the basis (in the sense of a narrow or broad understanding) and which planning practices are being discussed—does planning refer to the planning instruments, the planning object, or the planning context of a planning office, the city administration or a European funded project? Or, are planning practices in the United States, France or China meant? Planning practices are significantly shaped by the state and the respective understanding of the state in which they take place. As Kunzmann points out, you cannot rely on international theories or approaches developed in another country to understand one's own planning practice (Balducci and Bertolini 2007, p. 546). Specific administrative regulations, different distributions of responsibilities, laws or government systems fundamentally influence planning practice. The political context, the economic situation, can also be seen as having a significant influence on planning practice, since this context is responsible for the implementation of plans and measures (Tasan-Kok et al. 2016, p. 640). Other dynamics that affect planning practice are social changes such as population growth, technical and social innovations, specialisation, differentiation or complexity that create conflicts and enable new perceptions and new variants of possible courses of action (Bolan 1983, p. 29). Planning practice is therefore not absolute, but always to be defined, delimited and specified. How this is done, whether in terms of an interface with political processes, the activities and tasks of city planning officers, or decision-making in a democratic state, is fundamentally determined by the theoretical perspective that is adopted.

Implications for planning theory Both the consideration of planning practice as a research object and as a constitutive element of planning is reflected in planning theories. Planning practice as a research object contrasts planning practice with planning research and planning theories, mostly even dialectically with an insurmountable gap (Alexander 1997). Planning practice as a constitutive element of planning refers to the tautological relationship "planning is what planning does", of course not in the sense of writing emails or making phone calls, but understands ideal–typical and everyday behavioural norms as part of planning practice in addition to activities and actions. As agglomerated value-based actions, planning role understandings and self-images are thus introduced that relate either ideal–typical or everyday behavioural norms to specific planning understandings in a highly situation-related manner (Fox-Rogers and Murphy 2016, 75 ff.). It is an attempt to depict the plurality of forms of control, communication and knowledge inherent in planning and to give a simplified perspective of those who plan and what they do—the planning practice. Well-known examples are the planner as a helmsman or the planner as a moderator of divergent concerns. The ideal–typical role models serve to illustrate trends in planning theory (e.g. McLouglin 1970; Olesen 2018, 308 ff.) or the role of urban and spatial planning in a transforming understanding of the state (e.g. Fox-Rogers and Murphy 2016, 87 f.). Examples of this are the discussion about government and governance, a "turn" in the planning sciences through the change from a strong understanding of the state to a large number of actors in planning processes. A focus on everyday planning practices is based on the individual job description of individual planners (e.g. Healey 1992; Forester and Hoch 1996). The organisation of facts and arguments (Davoudi 2015, p. 318) as well as their evaluation and presentation (Dawkins 2016, p. 415) is therefore considered as important as the ability to persuade people to translate within the discourses or between different ones to navigate actors (Lothan in Tasan-Kok et al. 2016, 628 ff.). On the one hand, planning practice shows itself as a research object in the theoretical reflection of planning action and planning practices by specifically looking for specific role allocations. The challenges that such roles are strongly tied to the situation and can be assumed at the same time or in a chronological order are complex and conceptually poorly defined (Lamker 2016, p. 97). On the other hand, the understanding of planning practice as a self-constituting subject brings with it the additional difficulties that individual and situational actions have to be translated into generalised role understandings. Specifically, the question arises as to who this planner actually is, a clerk or a head of department of the urban planning office? And what is the role of planning when two planners take opposing positions to strategically influence political decisions (e.g. good-cop-bad-cop strategy). The equation that is as charming as it is simple, "planning is what planners do" is not able to adequately determine planning practice, since this is impossible as long as the epistemological object, namely planning, is not sufficiently determined, and as long as it is assumed that there is planning practice free of theory.

2.3 Diversity of Understandings of Planning—with Fundamental Differences

Necessity of definitions The understanding of planning is diverse, just like planning practices. This diversity enriches the field of spatial planning, but becomes a challenge in communicating and specificly describing what we imagine planning to be, which planning practice is assumed, which practices, activities and actions are to be understood by planning and to what extent this practice is generalizeable and how these practices condition, reflect, and specify our research approaches. The consequence of the diversity of planning practice is neither that there can be a generally valid and sufficient definition of planning practice, nor that the term planning practice is a term on its own. The great variety of understandings requires a precise definition and choice of terms with which planning is described in practice. Do we analyse planning practice as an act of public administration, as a compulsory task or as planning in private companies, which e.g. offer urban planning concepts? Is planning practice seen as a public service for the common good or as commissioned work? The precise definitional clarity of the terms of spatial planning or planning practice that is required here has far-reaching consequences.

Basic assumption of an identical understanding The term planning manifests a number of assumptions, which, however, are rarely made explicit in planning sciences—instead, the use of the word planning assumes a similar, if not identical, understanding of planning. There is a far-reaching difference between the naming of things (word) and the content, the substance of the word (concept) (Schönwandt 2008). The term must be explained by features and describes its content and meaning. This implies that what is meant, i.e. the concept, is the carrier of content, meaning and substance, not the word itself. The word is per se free of substance and meaning. Only the definition of a term makes it understandable and gives the language—the words—meaning, content and substance. What is understood by planning, for instance, or by planning practice, can vary (there is not one definition), but must be made explicit for mutual understanding (in order to at least be able to speak of the same term) (Schönwandt 2008). From a scientific-theoretical point of view, the basis of different understandings of a word is that we humans have no direct access to our environment, but can only take in stimuli with the help of our senses (sight, touch, hearing, smell and taste) and process them in our cognitive system. These differ depending on what is recorded and what is not, what is processed and how or with which further knowledge this is linked. Accordingly, terms cannot be used in a self-explanatory manner and your own recognition can only be a part of (individually constructed) reality. A look at the basic terms of spatial planning also reveals some interesting implicit definitions from a philosophy of science point of view.

Planning practice and theory? In the literature, the terms planning and planning practices are often used synonymously. However, the term planning practice is used specifically when it implicitly means the counterpart to planning theories and planning sciences. The title *There is no planning—only planning practices* (Alexander 2016) also points out that the term planning itself includes more than just the practices of planning. In other words, planning practice only reflects part of planning. This is important for the continuing discussion and highlights a fundamental scientific-philosophical distinction, namely to the relationship between (planning) practice and theory in the broadest sense. The prevailing position in planning sciences of practice without theory, i.e. the idea that we start from a practice to which we have direct access, is called positive realism (Barker and Kitcher 2014). This means that in this concept there is the one truth that can be described and that something of a universal practice exists beyond the cognitive capacities of any one of us. Accordingly, planning practice is an objective reality that can be represented objectively.

Positivism

Positivism sees it as necessary to recognise discoveries, which in turn should be considered knowledge, only on the basis of "positivist" findings and interpretations as such. They must therefore be traceable to sensory perceptible and verifiable findings. Post-positivism uses a critique of positivism to try to overcome it by identifying how objects are shaped by researchers (values, previous experiences, theories, hypotheses). Positivists in planning sciences face the challenge of using measurable and countable findings in planning as the basis of knowledge - exclusively measurable and countable findings.

Realism

Realism assumes that perception is always directed towards something existent and that things are as they appear to the respective viewer. Critical realism, on the other hand, shares the view of the existence of a real world that can be perceived by the senses, but this is not directly recognisable for human perception, and it remains unclear to what extent what is recognised by humans corresponds to the actual appearance. Critical realism appears as a perspective of planning sciences that may become increasingly important in the future. In particular, the combination of socially constructed realities and something that still exists, which is the basis of all rationalities in the sense of positivism, also corresponds to the planning orientation towards realities—the capacity to plan the world is highly dependent on our ability to agree on a reality.

Critical Realism This book and all further explanations are epistemologically in the tradition of critical realism, all the following explanations assume that the world is in principle recognisable, that we basically start from observation-independent existing objects about which we can produce knowledge (ontological perspective of realism). However, as humans we do not have direct access to these objects, but absorb reality through our sensory organs and our mental apparatus, i.e. a subjective processing of it (epistemological perspective of subjectivism). As a result, every consideration of planning practice is cognitively distorted and already filtered through theoretical glasses, sorted and not insignificantly shaped - in other words: Practice and theory are closely and inevitably linked. Rather than referring to a gap, theory and practice are more like two sides of a coin, one side does not exist without the other and together they make up the coin. Every practical planning consideration is therefore based on a basic theoretical attitude that is noticeable through our cognitive system. The distinction between whether we are all talking about the same practice anyway, since there is only one, or whether we first have to explain what understanding we have of the practice, i.e. the distinction between naive and critical realists, has far-reaching consequences for the understanding of planning and the way in which planning sciences are developed.

Unity of practice and theory On the epistemological basis of critical realism, we regard practice and theory as an inseparable unity that is mutually dependent and can in no way be treated as a dichotomy. It is impossible to look at planning practice without a theoretical perspective (see Fig. 2.2). Practice and theory are viewed from different perspectives that depict institutions, i.e. forms of living together that follow certain patterns. These perspectives can be summarised as science and planning; both planning sciences and planning practice reflect their characteristic perspectives on practice and theory. While the planning sciences in the traditional understanding of science analyse, research and try to understand practice in the form of empiricism (planning practice as a research object), they develop theories (further) and thus form a knowledge base of the professional field and the discipline. Planning, on the other hand (and this means all spatially effective actions and practices) is characterised by these practices and actions and is therefore sometimes used synonymously with the term planning practice; which is the subject of planning and is considered a constitutive element. The question that needs to be clarified here relates to the understanding of planning to be defined, for example planning in a narrow or broad understanding - a question that then has an essential effect on planning theories. Planning's perspective on theories can then be described as theorising, embedding the scientific planning theories in the locally specific context, reflecting on concrete planning actions and bases, as well as linking them to one's own experiences and translating them into the local context. The fundamental question here is which knowledge is actually assumed in planning, how this is obtained and what validity it has.

Underlying understanding of planning Basically, we try to make the respective understanding of planning explicit in the further explanations, especially when planning theories only become relevant when they are applied in the context of a narrow understanding of planning. A definitional classification of one's own understanding of planning can be explained relatively well with five answers. Starting with the distinction between a narrow and a broad understanding of planning:

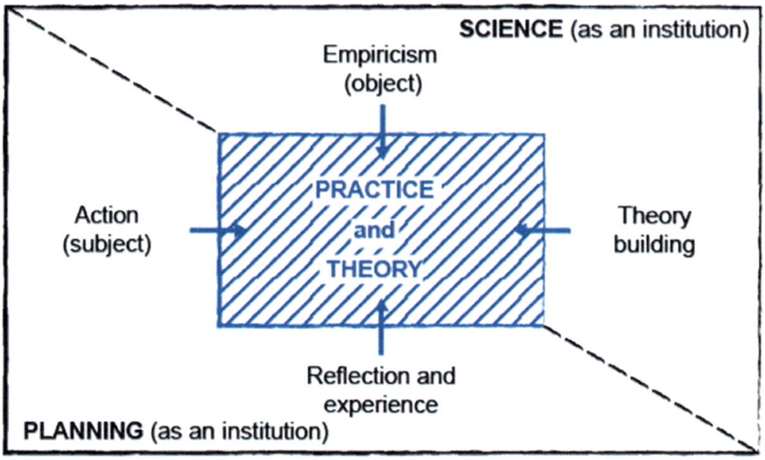

Fig. 2.2 Practice and theory as an inseparable unit in science and planning. (Own illustration)

1. Is state-led action the focus of my understanding of planning?
   However, since the term planning is not sufficiently defined by a narrow or broad understanding, a more detailed definition is necessary to clarify the term. Expanding on that:
2. What role do other actors and groups of actors play in decision-making? Do their perspectives have an informative, advisory or decision-relevant character?
3. What is the subject of the plan? Is it a plan, measures or the process of planning itself (epistemic subject)?
4. How is it planned and what mechanisms are behind it? What is intended by the determination of uses or by the selection of those involved?
5. How is the perspective of others dealt with in terms of the community, what significance do other rationalities have for spatial planning (basic ontological question)?

We fundamentally understand **spatial planning** as a social question of how we want to deal with the built and natural environment. Spatial planning uses different approaches to specifically influence future spatial development via physical and social control mechanisms, by (1) defining land uses and functions in the spatial context, and by (2) integrating different actor-specific perspectives, interests and control modes. A distinction can be made between spatial planning in a narrow sense as planning by the public sector and planning in a broad sense as spatially effective developments by a large number of actors. Various instruments are available for this, directly through formal planning instruments and indirectly through informal instruments and incentives, the steering potential of which depends on cooperation with other actors. The term actors is not necessarily reduced to another person, but stands for the introduction of a further perspective, a different view and cognitive evaluation of reality in the sense of critical realism. Spatial planning endeavours to understand different actor realities and rationalities in the conflict of objectives in order to shape spatial development.

References

Alexander ER (1997) A mile or a millimetre? Measuring the 'planning theory—practice gap. Environ Plann B Plann Des 24:3–6. https://doi.org/10.1068/b240003

Alexander ER (2016) There is no planning—only planning practices: notes for spatial planning theories. Plan Theory 15(1):91–103. https://doi.org/10.1177/1473095215594617

Balducci A, Bertolini L (2007) Reflecting on practice or reflecting with practice? Plan Theory Pract 8(4):532–555. https://doi.org/10.1080/14649350701664770

Bolan RS (1983) The structure of ethical choice in planning practice. J Plan Educ Res 3(1):23–34. https://doi.org/10.1177/0739456X8300300104

Brooks MP (2019) Planning theory for practitioners. Routledge, New York

Davoudi S (2015) Planning as practice of knowing. Plan Theory 14(3):316–331. https://doi.org/10.1177/1473095215575919

Dawkins CJ (2016) Preparing planners: the role of graduate planning education. J Plann Educ Res 36(4):414–426. https://doi.org/10.1177/0739456X15627193

Dobrucká L (2016) Reframing planning theory in terms of five categories of questions. Plan Theory 15(2):145–161. https://doi.org/10.1177/1473095214525392

Faludi A (1973) Planning theory. Pergamon Press, Oxford

Forester J (1982) Planning in the face of power. J Am Plann Assoc 48(1):67–80. https://doi.org/10.1080/01944368208976167

Forester J, Hoch C (1996) What planners do: power, politics, and persuasion. Adm Sci Q 41(4):719. https://doi.org/10.2307/2393873

Fox-Rogers L, Murphy E (2016) Self-perceptions of the role of the planner. Environ Plann B Plann Des 43(1):74–92. https://doi.org/10.1177/0265813515603860

Friedmann J (1987) Planning in the public domain. Princeton University Press, Princeton/New Jersey

Friedmann J, Hudson B (1974) Knowledge and action: a guide to planning theory. J Am Inst Plann 40(1):2–16. https://doi.org/10.1080/01944367408977442

Gunder M (2010) Planning as the ideology of (neoliberal) space. Plan Theor 9(4):298–314. https://doi.org/10.1177/1473095210368878

Hartmann T (2012) Wicked problems and clumsy solutions: planning as expectation management. Plan Theory 11(3):242–256

Healey P (1992) A planner's day: knowledge and action in communicative practice. J Am Plann Assoc 58(1):9–20. https://doi.org/10.1080/01944369208975531

Koch F (2015) The rules of the game and how to change them: urban planning between formal and informal practices. a colombian case study. Int Plann Stud 20(4):407–423. https://doi.org/10.1080/13563475.2015.1068685

Lamker CW (2016) Unsicherheit und Komplexität in Planungsprozessen [Uncertainty and complexity in planning processes]. TU Dortmund

McCann E (2016) Governing urbanism: urban governance studies 1.0, 2.0 and beyond. Urban Stud 54(2): 312–326. https://doi.org/10.1177/0042098016670046

McLouglin B (1970) Urban and regional planning: a systems approach. Faber and Faber, London

Olesen K (2018) Teaching planning theory as planner roles in urban planning education. High Educ Pedagogies 3(1):302–318. https://doi.org/10.1080/23752696.2018.1425098
Openshaw S, Veneris Y (2003) Numerical experiments with central place theory and spatial interaction modelling. Environ Plan A 35(8):1389–1403. https://doi.org/10.2478/mgr-2021-0019
Schönwandt W (2008) Planning in crisis? theoretical orientations for architecture and planning. Routledge, London
Selle K (2014) Über Bürgerbeteiligung hinaus: Stadtentwicklung als Gemeinschaftsaufgabe? Analysen und Konzepte [Beyond Citizen Participation: Urban development as a community task? Analyses and concepts]. Rohn, Lemgo (Urban Development Edition)
Tasan-Kok T, Bertolini L, Oliveira e Costa S, Lothan H, Carvalho H, Desmet M et al (2016) "Float like a butterfly, sting like a bee": giving voice to planning practitioners. Plann Theor Pract 17(4):621–651. https://doi.org/10.1080/14649357.2016.1225711
Throgmorton JA (1992) Planning as persuasive storytelling about the future: negotiating an electric power rate settlement in Illinois. J Plan Educ Res 12(1):17–31
Wiechmann T (2008) Planung und Adaption. Strategieentwicklung in Regionen, Organisationen und Netzwerken [Planning and adaptation. Strategy development in regions, organisations and networks]. Permission: Dresden, Univ., Habil.-Schr., 2007. Rohn, Dortmund
Wildavsky A (1973) If planning is everything, maybe it's nothing. Policy Sci 4(2):127–153

Planning Knowledge 3

Abstract

The epistemological question of knowledge of spatial planning can be expressed in knowledge of the present and the future. Planning theories reflect the ambivalence of the essential future orientation of planning knowledge and the importance of the normativity of this knowledge. Planning knowledge describes the fundamental establishing of planning approaches. A distinction can be made between descriptive, analytical, prescriptive and normative knowledge, however, decisive seems to be its normative knowledge content.

Keywords

Planning knowledge • Planning theories • Types of knowledge

Knowledge and action Dealing with planning knowledge is not an end in itself, but achieves e.g. its importance in planning processes: Political decisions are made on the basis of knowledge, which may have an impact on society as a whole. Accordingly, it is important to justify this knowledge and to outline what this knowledge is based on. Knowledge is a key component of planning, but subject to constant change over time. The focus of planning theories is primarily on actions or on knowledge that is directly linked to actions (Alexander 2001, 2016; Amin and Roberts 2008; Dalton 2007, e.g. data-related knowledge for making forecasts). In recent years, this has led to a trend towards evidence-based policies and practices (Davoudi 2015, p. 316). Different forms of knowledge play an important role here, since the underlying knowledge determines actions and thus the results of planning processes. Whether technical expertise, the use of forecasts to check actions for their future viability or the application of analytical methods. This knowledge defines the framework in which planning can exist in the first place. At the same time, it can be assumed that actions and power structures in turn influence knowledge. Knowledge and the generation of knowledge have a serious impact on the lives of a large number of people, since spatial planning structures and designs

M. Levin-Keitel and L. Behrend, *The Topology of Planning Theories*, The Urban Book Series, https://doi.org/10.1007/978-3-031-37857-7_3

(living) spaces. This situation gives rise to a couple of questions that should be considered here: What is knowledge? What is considered knowledge in planning sciences?

3.1 Knowledge About Knowledge in General

Knowledge in the Classical Sense The classical understanding of knowledge was written down by Roderick Chisholm (1957: 16):

"´*a* knows that *p*´ means:

(i) *a* accepts *p*,
(ii) *a* has adequate evidence for *p*,
and
(iii) *p* is true.

According to this, knowledge is identical with justified true belief. This view can be traced in a less formal presentation e. g. by Plato, Aristotle, René Descartes, Immanuel Kant, Edmund Husserl or the empiricists in general (Davoudi 2015, p. 3) but has been much criticised and even refuted by Edmund Gettier (1963). Pursuing the question of how this perspective on knowledge, which is still influential today despite the fundamental criticism of it, came into being, represents the basis for a reflection on planning knowledge and explains different epistemological perspectives of planning as well. A detailed review and classification of knowledge about knowledge seems necessary.

Early Forms of Explanation The search for explanations, the quest for knowledge, is one of man's oldest cognitive endeavours. A number of different approaches have emerged. The myth, is the earliest form of explanation of the world (Russell 2009). There is no distinction between the real and the additional thought, but these are woven together in a narrative. Empirical knowledge is directly linked to an animistic idea, since it was assumed that everything—every thing—has a potential soul. Accordingly, there is no theoretical justification, but the validity of the narrative stems from the belief that it is true (Russell 2009). There are no experts on lore and "fiction"; rather, it is a collective effort, developed by individuals, but preserved and passed on—orally—by society as a whole. Narratives are self-contained and incompatible with others. Myth is the classic worldview of archaic societies, even if contemporary esoteric groups prove that such worldviews still exist (Russell 2009). Religion replaces myth as a system of interpreting the world with a theological one. This form of explanation differs from myth in two main ways: On the one hand, the collective performance is replaced by a group of experts with authority to interpret. On the other hand, there is a close connection to the ruling class in which it is given legitimacy (Russell 2009). The theological justification develops a coded system consisting of rites, prohibitions and dogmas, and through the institutionalised religion represents an extremely consistent social and cognitive system (Russell 2009). A justification of cognition and knowledge is based on the assumption of enlightenment by the respective god or gods and accordingly leads back to the theory in a circular manner. The emergence of civil society led to a revolution in the development of theories (Sahakian 1968). This differs from the agrarian society in that it no longer secures its livelihood through primary production, but through crafts, trade and services. This presupposes the availability of food (e.g., through procurement by slaves or other means) and the fact that an increase in productivity frees a larger part of the population from work in primary production. It also assumes that the population lives in larger numbers—mostly in cities—and is not under the tutelage of a ruling class. Only this constellation makes it possible to overcome myth and religion (Solomon and Higgins 1996). The Greek culture is regarded as the first historical constellation in which this happened. In this cultural space, philosophy emerged, which represented a completely new way of reflecting in the form of a theoretical approach to the world in a systematic way (Russell 2009). The pre-Socratics and Sophists are not discussed in detail, but Socrates, Plato and Aristotle cannot be left out, since their questions and lines of thought still have a decisive influence today and determine many theories through their epistemological classification.

Socrates and true knowledge in every human being Socrates' worldview aims to reconcile action and knowledge—in this way actions can be normatively justified. For Socrates, right knowledge leads to right action (Russell 2009) (as we have already seen, this is not necessarily the case in general as well as in planning). But, how do you get to the right knowledge? To do this, Socrates uses "Maieutik" (the art of midwifery), which brings forth the knowledge buried in every human being and is intended to make it accessible. In a first step, false knowledge must be exposed through targeted and persistent questions in order to create the starting point for the development of true knowledge on the basis of the insight that nothing is known for certain. "True knowledge" is already present in every human being and only needs to be uncovered. So, for Socrates it is not so much the knowledge that is a problem as the finding of the knowledge. For him, the key lies in a critique of "wrong knowledge" (Russell 2009).

Plato and the idea of the good Plato, student of Socrates, detaches himself from the worldly focus and instead develops a system of the objectively good and true (Sahakian 1968). He builds on a dualism by dividing reality into the empirical, which is changeable and transient, and the fundamental, which holds the world together in its entirety. He calls these ideas. The idea of the good is supreme. For Plato there is no real knowledge in the empirical world, only opinions that can be more or less correct. He states this in his famous allegory of the cave. True knowledge, on the other hand, is only possible in the realm of ideas and therefore not through sensory experience. For Plato, knowledge is to be understood as recognition, since he assumes that the soul has already grasped things at an earlier stage (Russell 2009). The method of recognition is thinking, which must be trained and freed from error, since the perception of something is equivalent to the knowledge of truth (Russell 2009). Since true knowledge can only be found in the realm of ideas, Plato is also referred to as an "idealist".

Aristotle's syllogistic Aristotle, a student of Plato, proceeds epistemologically differently from Plato in many respects. Although he is a "Platonist" with regard to the dual understanding of the world—the order of the empirical world through a non-material background—reality results for him primarily from "material" and formative principles (Russell 2009). For Aristotle, "being" is movement and behind every movement there must be a moving force. In this way, a new concept is established as a driving principle: causality. Aristotle's ontology of being is a justification that does not require external reasons (as in a religion). Due to this, all being is a part of the logically ordered cosmos—therefore the comprehensibility through thinking is the possibility to recognise this (Russell 2009). From this, Aristotle founds his own world of knowledge, which is self-contained and connects being and thinking as well as empirical and transcendent reality (Russell 2009). For him, thinking is divided into three different elements: Begriffe, Urteile und Schlüsse (terms, judgments and conclusions) (Schülein and Reitze 2012). The definition is the ideal term. The essentials are contained here precisely, completely and unambiguously. It does this by specifying the similarities that something shares with something else and the specifics that make it different from everything else. However, terms only designate and say nothing about what the case is. Statements about reality can only be made through their context and only then can it be decided whether they are true or not. The statement about a connection is a judgment (Russell 2009). With the verdict, the truth is reached, which Aristotle calls conformity with reality and logical coherence. The conclusion now presents possible consequences. The syllogism is the form of inference that enables the transfer of a truth found through judgments to another state of affairs. „Der Syllogismus ist eine Gedankenverbindung, in der, wenn etwas gesetzt ist, etwas anderes als das Gesetzte notwendig folgt, und zwar dadurch, daß das Gesetzte ist " ["The syllogism is a combination of ideas in which, when something is posited, something other than what is posited necessarily follows, and that is because what is posited is"] (own translation from Aristotle 1921, p. 24b18). It therefore allows generalisations and the transfer or a system of truths—i.e. a theory (Russell 2009). With syllogistic, Aristotle creates a verifiable procedure that normalises knowledge in the sense of a functional logic. „Im Begriff wird Wirklichkeit logisch operationalisiert, im Urteil erfolgt eine Überprüfung des Wahrheitsgehalts, im Schluss die Systematisierung und Generalisierung " ["Reality is logically operationalised in the concept, the veracity is checked in the judgment, and finally the systematisation and generalisation"] (own translation from Schülein and Reitze 2012, p. 51). Aristotle is therefore regarded as the founder of modern scientific epistemology.

Implications for planning theories These three philosophers have had a significant impact on the development of (especially western) philosophy and thinking (in general, through colonialization) to this day. Although philosophy was often used for theological purposes in the following centuries (as by scholasticism), a variety of discourses arose that had a great influence on later debates. Emblematic of this is the problem of universals. Here the question is discussed whether general concepts exist before all reality (as in Plato's idealism) or in many details (according to the realists) or are only ex-post constructions and designations of our mind (according to the nominalists) (Russell 2009). A clearly epistemological debate that already includes the positions of idealism, realism and constructivism which are fundamental for planning sciences and planning theories: While planning knowledge in the sense of idealism can only be found in the realm of ideas and includes more ideal–typical planning processes, planning knowledge in realism is based on material principles such as land use as a result of planning processes. In constructivism, where all knowledge is considered socially constructed, there are many different knowledge claims that want to be considered.

The Middle Ages as a turning point In the late Middle Ages, the general conditions of feudal society changed and a new type of society emerged. As Europe opened up to the outside world, trade and crafts as well as new cities developed, which gained a high degree of autonomy. As a result, a bourgeoisie emerged that slowly emancipated itself from paternalism, both intellectually and politically (Solomon and Higgins 1996). This bourgeoisie developed a new ideal of education and the "researcher" emerged: a committed private individual, a bourgeois expert. Researchers were no longer clerics and members of the universities but a bourgeois, public discourse emerged. Salons and academies took the place of the university, which had previously been purely clerical. The researchers were amateurs, but the methodology became more professional. The results were more and more systematic collections, experiments and interpretations of empirical (Russell 2009). Research was often pursued as a hobby alongside one's actual job and it was quite common to devote oneself to different topics. However, this unity of reflection dissolved as knowledge increased—the discourses diversified (Russell 2009). This development led to an increasing need to think about cognition itself and, above all, to answer the question of how knowledge is acquired. There were basically two answers to this question: Rationalism and empiricism

Rationalism

Empiricism on the other hand, emphasizes the importance of real experience for cognition. It ties in with nominalist thinking and emphasizes sensory contact with the world. Francis Bacon is considered his pioneer (Russell 2009). In his main work Novum Organum (translated as "New Tool"), he strives for a methodical approach or instrument for all sciences (Bacon 1990). The highest science thereby watches over the cognitive ability of the other disciplines by criticizing them. True knowledge is only possible if we free ourselves from prejudices, which Bacon calls "idola". He recognizes four of them. The "idola tribus" (idols of the tribe) arise from belonging to the human species. Since we can grasp the reality only with the means of the intellect and the senses, always a distorted picture of the reality develops (Russell 2009). The "idola specus" (idols of the cave) are individual prejudices that arise from predisposition, upbringing, or habits and inclinations (Russell 2009). The "idola fori" (idols of the market) are language problems that, following the problem of universals take a nominalist position. They arise from the fact that the "names of things" are not connected with their logic, but arise in social intercourse (on the market, so to speak) (Russell 2009). Finally, there are the "idola theatri" (idols of the theatre). These emerge in philosophical schools when theories become increasingly entrenched and dogmatic - they become institutions and dominate thought (Russell 2009). For Bacon, only the interplay of empirical experience and organizing mind leads to a true picture of nature (Russell 2009). In this respect, his concept of truth resembles utilitarianism and contemporary neo-constructivism themes. Russell (2009) sees Bacon's work as a strategy paper with scientific-political intent; with the goal of establishing a critical community. With his induction, he also represents how research happens in the first place and the attempt to insert new observations into existing theories through discursive consensus.

Rationalism assumes that knowledge does not result solely from empirical facts, but is the result of intellectual achievements, the "ratio" (Russell 2009). Early and influential representatives of this trend are René Descartes, Gottfried Wilhelm Leibniz and Baruch Spinoza. In particular, Descartes' methodological doubts are decisive for this trend. He starts from the undeniable possibility of error, which arises from the fact that everything is or can be empirical deception (Descartes 2009, p. 20). The only definitive point of reference for knowledge is the mentally active ego (res cogitans), which deals with the material world (res extensa) (Descartes 2009, p. 30). The world is captured in ideas (Russell 2009), knowledge and acquisition of knowledge follow one rule: „Es muß das Ziel der wissenschaftlichen Bestrebungen sein, den Geist so zu lenken, daß er über alle sich ihm darbietenden Gegenstände begründete und wahre Urteile fällt " ["It must be the goal of scientific endeavour to direct the mind so that it makes reasoned and true judgments about all objects presented to it"] (Descartes 1960, p. 21). Knowledge can be acquired in this area through intuition ("above all doubt comprehension of a pure and perceptive mind") (Descartes 1960, p. 26, is what Descartes means by this) and deduction (logical deductions from pre-existing knowledge). Mathematics plays a special role here, since it is used to methodically record knowledge (Russell 2009). Knowledge thus becomes the achievement of the "res cogitans" that confronts the world—a separation of subject and object resulting from the division into mind and matter. Descartes' role in philosophy and thinking can hardly be overestimated, since the modernisation of the same arose in this way. It was only through this that the disposal of the world of objects (necessary for planning) and the dissolution of the unity of the world came into being (Russell 2009). The increasing mechanisation of thinking also begins with Descartes—he partially breaks away from theological thinking, remains idealistic in the platonic sense, but ties knowledge more strongly than before to a methodology (Descartes 2009, p. 79 ff.; Russell 2009).

Methods of knowledge The discourse on rationalism and empiricism is significant because of the new confrontations with methods of knowledge (see Fig. 3.1). Of course, this also applies to the planning sciences and the development of knowledge. Until now, the search for knowledge has been a cognitive enterprise. Rationalism attempted for the first time to establish theoretical concepts of contact with reality and thus established a strand of modern epistemology (Russell 2009).

Rationalism uses reason as a mode of cognition, the object of cognition is described as dependent on consciousness, in other words cognition is first formed with one's consciousness (a priori) before it can be empirically demonstrated. In terms of planning knowledge, this means that abstract concepts and ideals are first presumed, which are then to be empirically proven. The other strand is based on empiricism, which focuses on the practice of knowledge and fixes it on methodically controlled results (Russell 2009). Empiricism takes knowledge inductively (through the senses) and assumes an object independent of consciousness, which can be recognised empirically. With regard to planning knowledge, empirical knowledge is limited to what can be measured and carried out in experiments. The

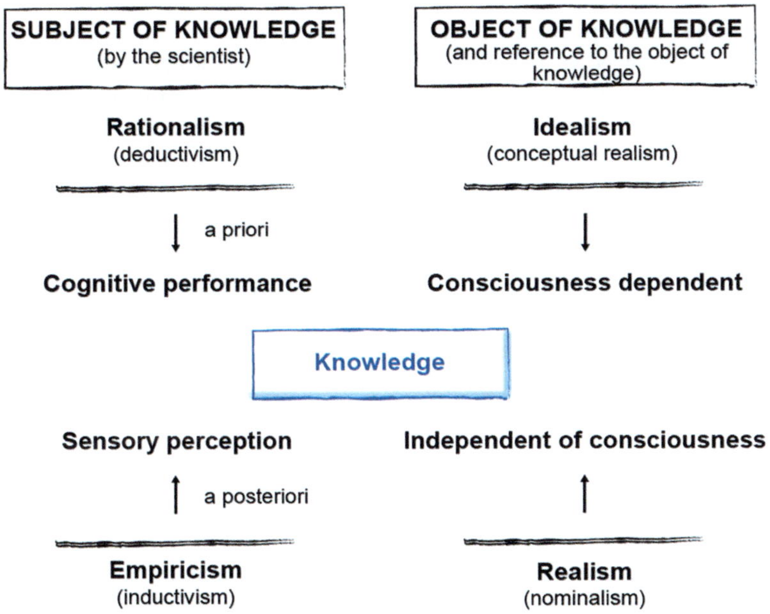

Fig. 3.1 Epistemological basic positions (Own illustration)

use of observation and experiment was methodologically groundbreaking. Observation is first and foremost an active perception of reality, but always selective and constitutive with regard to the theoretical prerequisite (the unity of theory and practice). It is important whether the observation is subject to everyday consciousness or whether it is controlled. Context-independent knowledge can only be gained on the basis of controlled observations—and only in this way can they form the basis for empirically founded theoretical statements (Sahakian 1986). The experiment represents the step beyond observation. On this basis, new experiences should be collected and hypotheses checked—reality is actively processed here (Russell 2009). Accordingly, prediction and criticism are the most important tools of the experiment that can be transferred to the theoretical framework (Barker and Kitcher 2014). In addition, the metricization of data has led to an expansion of qualitative terms to include increasing quantification. By counting and measuring, facts can be defined more concretely and determined more precisely (Russell 2009). This metrication, combined with observation and experiment, is characteristic of the empirical sciences.

Kant's transcendental philosophy as an intermediate form Immanuel Kant takes the view that no theory can arise from empirical experience (Russell 2009). He tries to combine the positions of rationalism and empiricism into a new foundation—a transcendental philosophical one. Kant's goal is to reconcile empirical and rationalistic assumptions (Russell 2009). He starts from the question whether synthetic a priori judgments are possible (judgments that are independent of experience but not inherent in concepts). So they are a kind of paradox. Kant focuses on both experience and thought. Not just one of the two: Experience without concepts is blind; Concepts without experience are empty. What is decisive for Kant in relation to knowledge, however, is whether there are structures of thought that are independent of experience. Kant proceeds as follows: Human consciousness has certain forms of thinking that make experience possible. On the one hand the forms of perception (the order of things in space and time) and on the other hand the categories of understanding (such as causality) (Kant 1781/1787 KrV A/B). However, every experience is preceded by other basic concepts of reason—the mental instruments that make it possible to grasp reality in the first place (Kant 1781/1787 KrV A/B). The point is that every experience is based on its principle. Recognition is therefore not passive perception, but an active process, since here the basic ability to recognise is combined with the appropriate means (Kant 1781/1787 KrV A/B). The knowing subject thus uses his transcendental cognitive faculty according to rational methods. For Kant, cognition is dependent on both reason and methodology (Kant 1781/1787 KrV A/B). Central to this is the positioning of theory and object—for Kant, the object is dependent on the theory. From this perspective comes the name of the Copernican turn in thinking, which Kant postulates on the basis of a comparison. Of the objects of experience only what the mind can comprehend can be grasped, only non-empirical facts can be grasped by the mind. However, he emphasises the possibility of methodological control, since experience and knowledge function reciprocally and the activity of knowledge must be checked against reality. According to this, however, objects cannot be grasped in themselves, since they are always formatted by the performance of the mind (Russell 2009). The *thing in itself* is not attainable, but only what the mind makes of it. So there is an essential difference between the real object and the object reflected by a subject. In a way, Kant is a forerunner of constructivism, which makes it clear that theory is always just a reconstruction of an object and never a representation of it. What these fundamental differences mean for spatial planning and their findings is exemplified in Sect. 3.2.

Criticism of rationalism and empiricism (also in planning sciences) The view shared by rationalism and empiricism that an ultimate truth can be traced is criticised from different positions. Sceptics doubt that there is any fundamental truth at all (Comesaña and Klein 2019) and pragmatists assume that truth is proven in the process and that there is accordingly no fundamental truth (Legg and Hookway 2020). In the planning sciences, Davoudi (2015, p. 3 f.) lists this form of criticism as criticism of the analysis of epistemic knowledge. This criticism introduces two further forms of criticism. One relates to the sources of epistemic knowledge. She shows that both the tradition of empiricism and that of rationalism belong to the naturalistic research tradition (Davoudi 2015, p. 4). This is criticised by hermeneutics, since it assumes that the social world can only be understood from the inside and not from the outside (Guignon 1991, p. 84 ff.). For planning, this criticism means that the perceptions of individual planners and the social rules that guide and restrict the actions of the planners must be taken more into account (Davoudi 2015, p. 4). Social science currents that emphasise this include constructivism, postmodernism, and critical realism. They agree that knowledge is socially and historically constructed (Davoudi 2015, p. 4). Davoudi's second critique relates to the types of epistemic knowledge. Here she criticises that only one kind of knowledge is considered valid: Those of the naturalistic tradition. Other types are excluded (Davoudi 2015, p. 5).

Other basic terms Let's now try to bring together basic terms that emerge from what has been said before and dare to look beyond them.

A priori and a posteriori

Knowledge can be divided into a priori (knowledge independent of experience; such as logical, or mathematical, knowledge) and a posteriori (knowledge based on empirical perceptions and experiences; such as scientific etc. knowledge). Rationalism only recognises a priori knowledge as the basis of our knowledge, while empiricism is based on a posteriori knowledge. A variant of the epistemological fundamentalism just presented is the transcendental philosophy, which assumes that there is knowledge a priori as well as a posteriori.

Naturalistic fallacy
A naturalistic fallacy is the transference of the property *good* to a relation or property of a descriptive, metaphysical, or natural type. Naturalistic fallacies can also be diagnosed in planning, in particular due to their future orientation: Because from a statement that describes how it is (descriptive statement), you cannot derive a statement of how it should be (normative statement). Or even clearer: *There is no direct link from an is to an ought!* This means that even a detailed analysis of a status quo in planning always contains many options (and potential recommendations) for planning the future.

Constructivism
Constructivism is an epistemological position that focuses on the construction of the recognised object by the subject. As already shown, this position is similar to the nominalist one in the problem of universals. In the end, different constructs of reality stand side by side, no final truth can be determined.

Contextualism
Contextualism makes knowledge partially dependent on historical and situational contexts (Detel 2014, p. 73). Accordingly, no fundamental knowledge is recognized. Instead, opinions can represent knowledge in certain contexts and not in others (Detel 2014, p. 74). Contextualism is widespread in planning, examples are the social context, the political context or even the spatial context. Contexts serve as a kind of black box that makes further explanations unnecessary and underlines the uniqueness of planning processes.

3.2 Knowledge in Spatial Planning

Knowledge and science Sciences are considered to be systematised knowledge in the form of disciplines. Knowledge, as the heart of science, must meet far-reaching knowledge claims, which in turn are manifested by science and its function as a *gatekeeper*. The claims to knowledge of both the disciplines and science itself extend well beyond objects of sensory perception alone. For instance, these knowledge claims can be objects from the distant past, such as questions of evolutionary biology, or objects that are not part of our direct experience, such as atoms or molecules. However, the claim of science goes even further, since it not only wants to be descriptive, but also provides explanations. Accordingly, science develops an explanatory character (e.g. by finding and describing natural laws). Science in the classical understanding refers to natural sciences. Three main types are distinguished when speaking of science: Science, Humanities and Social Sciences

(Barker and Kitcher 2014). But where in this triad can spatial planning be found? Like mathematics or engineering, it seems to have no place. Mathematics is often given an additional place outside of the three mentioned, and the same goes for engineering (and thus spatial planning in the broadest sense). This already shows that the question of locating planning sciences is not easy to answer. Science is multifaceted and extremely complex. Nevertheless, three common ideas about science can be identified. (1) The technical-scientific idea, largely shaped by Bacon, uses science as a reliable means of accumulating useful knowledge. Observations are used to derive generalisations from individual cases and to draw up general conclusions. The goal of this accumulation of knowledge is to gain control over nature and use science as a tool to generate human benefit. (2) Differentiating from this is Karl Popper's (1963) view. This considers the observation as a misguided starting point, since simple observation is not possible if the observer does not know what to observe. Instead, it is the task of the scientists to come up with hypotheses that are as risky as possible, which are then to be disproved by experiments. Hypotheses, no matter how certain they may seem, are never fully accepted. Ideally, these are provisional truths that help people on their path of knowledge. (3) The third idea of science, and the most common one today, portrays it as the epitome of rationality. Science is therefore considered the best way to understand the meaning of the world through the human mind. Even if scientific knowledge is never definitive, there are objectively measurable connections between evidence and theories or hypotheses. Therefore, important scientific findings, such as e.g. the knowledge of the laws of chemical reactions, can be regarded as objectively established. They are not immune to revision, but as close to certain knowledge as we can get. Science as a whole aims at a systematically unified picture of nature and the last decades are seen as a big step in this direction. This third understanding of science also underlies spatial planning, which often assumes fundamental rationalities in planning processes and questions. However, the question arises as to which forms of knowledge form the basis for planning processes.

Multiple Forms of Knowledge Let us now address the question of how knowledge is represented in planning (theory). Yvonne Rydin (2007, p. 53) distinguishes knowledge from information and data through a causal relationship. It shows why knowledge is so relevant for planning: "Since planning seeks to create specific impacts, planners need to understand how such impacts follow from specific planning actions; they need to understand the causal relationships between action and impact" (Rydin 2007, p. 53). Following Sandercock (1998), Rydin also sees planning as an activity that has its origins in the "modernist conception of society" (Rydin 2007, p. 53). Accordingly, the intention of planning is to contribute to or produce a social good through intervention. Knowledge plays a special role here, as Rydin has illustrated using the above definitions. Knowledge should no longer be viewed as a unifying category (Evans and Marvin 2006), knowledge is inherently multiple, since it contains different claims to the representation of reality and represents different forms of knowledge (Sandercock 1998). This perspective on knowledge stands in contrast to the positivist claim that assumes the existence of a single truth (Rydin 2007, p. 54). The increasing repression of the modernist view of knowledge prompts Sandercock (1998, p. 76) to call for "an epistemology of multiplicity". This includes certain forms of knowledge: "through dialogue, from experience, from local knowledge, by learning to read symbolic and non-verbal evidence; and through contemplative or appreciative knowledge" (Rydin 2007, p. 55). Rydin (2007, p. 55) even sees the task of planning itself in processing and dealing with the various forms of knowledge.

Different knowledge claims and expansion of those who know Knowledge is therefore no longer just the domain of experts, but is being expanded to various actors and localities (e.g. no longer just the university) (Rydin 2007, p. 54). Different forms of knowledge emerge from these different sources of knowledge (Rydin 2007, p. 54). Assuming multiple forms of knowledge, the difficulty of translation between different forms and actors arises—especially between lay and expert knowledge (Evans and Marvin 2006). Innes (2004) also shares this view and points out that in a public discourse consensus-building is achieved less through better arguments than through better storytelling. It is important to point out that it makes a significant difference whether different voices have to be heard or whether different knowledge has to be implemented, i.e. has a guiding character (Rydin 2007, p. 56). So there are different requirements in a planning process, each of which has its justification, but is fundamentally different. With the "communicative turn" a mixture of these demands has arisen at the same time, in the constructivist sense, in order to be able to make the best possible and socially robust decisions. In particular, there is a tendency to declare a variety of claims as knowledge at the local level, although they are not always claims to knowledge (Sandercock 1998). Rydin (2007, p. 56) sees this formulation primarily as an attempt to increase the importance or status of the local level. Knowledge claims are also differentiated within themselves, e.g. communities can have historical knowledge of their city's cultural heritage while maintaining emotional connections to it. Both aspects are important for the overall claim of the community, but they form different claims to knowledge (Rydin 2007, p. 57). To do this, the voices of all actors must be recognised and listened to by the planners (Rydin 2007, p. 57). Rydin (2007, p. 58) sees this as part of the Habermasian tradition, in which the illocutionary speech act, i.e. the action performed through language, must also meet validity requirements. Rydin (2007, p. 58) sees the planning system as a series of arenas in which the different claims to knowledge meet. Planners not only have to be responsible for the procedural aspects, but are also actively involved by co-producing knowledge "through testing and recognising knowledge claims" (Rydin 2007, p. 58). This includes opening up the discourse for the different voices, but also summarising and limiting to knowledge claims.

Knowledge perspectives in other disciplines
Outside of philosophy and planning theory, debates about knowledge have attracted just as much attention. The sociology of knowledge deals in a narrower understanding with the socio-cultural influence on the formation of opinions and beliefs, in a broader understanding also with the influences on knowledge (Merton 1937, p. 493). Max Scheler and Karl Mannheim in particular have dealt extensively with the sociological aspects of knowledge. In addition to these two, Michael Polanyi should also be mentioned. His discovery of tacit knowledge is particularly significant. "Tacit knowledge is considered as personal knowledge that is difficult to express, formalise or share and exists in an intangible format" (Chugh et al. 2015, p. 129). Polanyi (1966, p. 4) himself summarises the concept of "tacit knowledge" as follows: "We can know more than we can tell." Recently, Michel Foucault (1971) has also made important contributions to the field of the sociology of knowledge (and far beyond). These have also been widely received in planning theory. In particular, Foucault's (1977) perspective on power, is used by authors such as Forester (1989) or Flyvbjerg (1998) as a starting point for their discussions of planning practice.

Types of knowledge in planning Three authors in particular have dealt with the concept of knowledge in planning: Yvonne Rydin (2007), Heather Campbell (2012) und Simin Davoudi (2015). All authors define different types of knowledge, each of which is not value-free and contains more or less causal relationships (Table 3.1). Rydin (2007, p. 63 f.) identifies four knowledge claims: empirical, processual, predictive and normative knowledge. Empirical knowledge is expert assessments or lay knowledge, which planners must check or collect themselves (Rydin 2007, p. 63). Process knowledge includes the planning organisation and the socio-economic context of the planning process. Predicative knowledge includes the future and how it is planned and is dedicated to future trends and developments, just like normative knowledge that deals with the goals of planning. Campbell sees four central questions as they arise in the field of planning and politics: What is going on here? Why is it like that? What to do? What ought to be done? Each of these questions is the subject of a different kind of knowledge. As in Rydin's identified types of knowledge, the first two questions aim at the current state, in a descriptive sense and in the sense of an analysis of why something is the way it is.

The goal orientation results from the prescriptive and normative future perspective—from the last two questions (Campbell 2012, p. 138). In doing so, Campbell places the relationship between is and ought at the centre of her further considerations. Davoudi understands planning as the practice of knowledge—knowledge is not something that planners have, but something that planners do (Davoudi 2015, p. 1). „... it is more useful to think about planning as practice of knowing that involves knowing what, knowing how, knowing to what end and doing " (Davoudi

Table 3.1 Types of knowledge in planning (Own illustration)

	Rydin (2007)	Campbell (2012)	Davoudi (2015)
Types	• Empirical knowledge • Procedural knowledge • Predictive knowledge • Normative knowledge	• What is going on here? (descriptive) • Why is it like that? (analytical, explanatory) • What to do? (descriptive) • What ought to be done? (normative)	• Knowing what (episteme) • Knowing how (technical) • Knowing to what end • Doing (phronesis)
Peculiarity	Distinguishing between knowledge about the present state (empirical and procedural) and the future (predictive and normative)	Ask how knowledge can be translated into action; relationship between is and ought	Planning knowledge lies in the actions, not just in books or data

2015, p. 1). She refers to Aristotle's *episteme* (rational understanding of planning as true opinion), *techne* (questions of skills and craft) and *phronesis* (correct practical judgment and desire). Planners become virtuous or wise (in an Aristotelian sense) when they achieve this correspondence, which lies in practical action and not in a priori knowing what to do. What planners know also lies in the actions and not only in books or data collections (Davoudi 2015, p. 6).

Planning knowledge in general What all three approaches by Rydin, Campbell and Davoudi have in common is that they define four types of knowledge: Knowledge from information and facts (descriptive), knowledge from the analysis of these facts (analytical), knowledge about the future (prescriptive) and knowledge about the ought (normative) (see Fig. 3.2): They recognise a form of knowledge that deals with the collection of information and facts. For Rydin (2007) this is empirical knowledge, for Campbell (2012) descriptive knowledge, and for Davoudi (2015) it is most closely related to knowing what, although here it is least clear-cut. In addition, everyone has an understanding of knowledge in the sense of an analytical consideration of the facts at hand. For Rydin (2007) this falls under process knowledge, for Campbell (2012) under analytical and for Davoudi (2015) it is less clearly assignable and partly plays a role in both "knowing what" and "knowing how". In addition, the approach of prescriptive knowledge can be found in all of them. For Rydin (2007) in predictive knowledge, for Campbell (2012) in prescriptive knowledge and for Davoudi (2015) again both in "knowing how" and in "doing". In addition, all three identify normative knowledge. Rydin (2007) in the eponymous, as well as Campbell (2012) and Davoudi (2015) in "knowing to what end". Finally, Davoudi (2015) defines "practical judgement", which is less a specific form of knowledge than virtuous action, in the Aristotelian tradition.

Implications for planning theories The different perspectives on the knowledge of planning impressively show the field of tension in which planning acts. What constitutes planning varies depending on the position towards knowledge and the

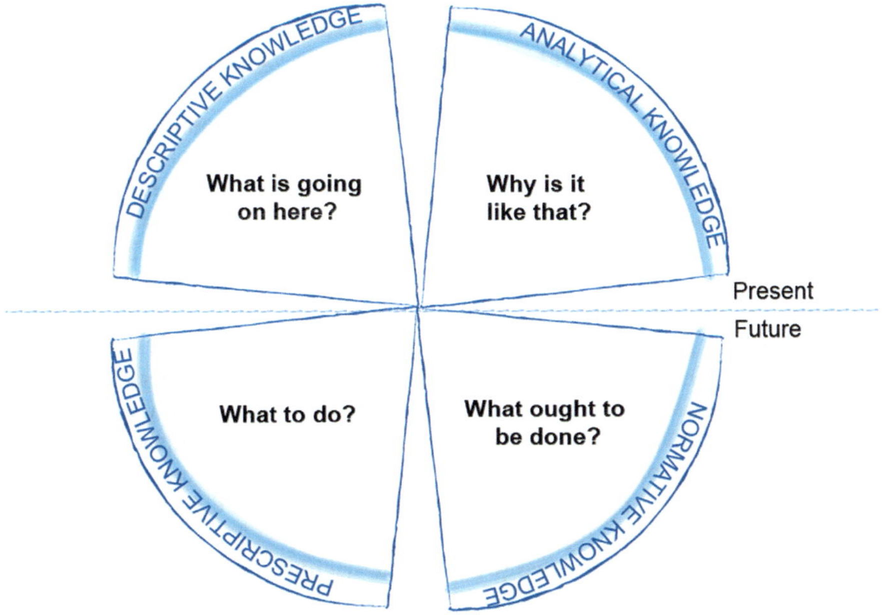

Fig. 3.2 Types of knowledge in planning. (Based on Rydin 2007; Campbell 2012; Davoudi 2015, own illustration)

knowledge claims to current realities or future ideas. Some further conclusions for knowledge in planning can be drawn from the types of knowledge and their critical reflection:

1. **Knowledge of present or future:** According to Campbell, a key misunderstanding between practitioners and academics in planning can be identified from the distinction between knowledge about the current state and the future. Practitioners expect improvements in relation to the present—it seems more useful and helpful for daily work when this knowledge formulates what is happening and what should be done. Knowledge about the future, on the other hand, is much more fundamental and will transform current practices—but there is little research on this (Campbell 2012, p. 138). This also has to do with the difficulty of translating such insights into practice.
2. **Being in the post-positivist sense does not exist:** The relationship between what is and what should be can be deepened from the discrepancies between current analysis and future goals. David Humes already formulated the concept of the naturalistic fallacy here, that an ought cannot be derived directly from what is. Campbell (2012, p. 139) comes back to this and explains why both the positivist and the post-positivist view do not allow inferring from a *what is* to *ought to be*. The first assumes that no *ought* can be derived from *what is*, the second that there is no *what is* and accordingly no *what ought to be*. From the

post-positivist perspective, knowledge is socially constructed and accordingly there is not and should not be a single *what* (Campbell 2012, p. 139). Only an individual view of the world exists and therefore a variety of forms of knowledge need to be acknowledged. This is relevant because in recent years there has been an increase in evidence-based policy and, in general, more reference to facts and information in decision-making (Davoudi 2015, p. 1; ODPM 2004, p. 39 f.). The problem, however, is that knowledge of a lower life expectancy in a district, for instance, does not automatically point out what should be done about it. This shows that while there may be a scientific evidence base, it is not necessarily linked to policy decisions about what to do (Lindblom and Cohen 1979). The rational approach to planning tries a positivist position, in which *what is* seems to be separated from *what ought to be*. According to this, politicians decide on the *what ought to be* as the goal, while planners provide the means to achieve it. So in this sense planners do not deal with values. However, Faludi and Waterhout (2006, p. 9) have already stated: This is not possible because the decision of what counts as evidence is already value-laden. Accordingly, this positivist position is untenable. It follows that decisions about what to do are, and must be, highly contested.

3. **Equivalent claims to knowledge negate an orientation towards the common good:** According to this, different claims to knowledge already exist in the description of the status quo, the current knowledge. These are often presented in planning with the approaches according to the *communicative turn* and the normative claim to offer everyone the opportunity to participate. The aim is to curtail government power by involving as many voices as possible (Campbell 2012, p. 140). Reference is made here briefly to the ontological change from a narrow to a broad understanding of planning (see Chap. 2). The fundamental problem here is that these many different claims to knowledge must be prioritised among themselves on the one hand, and on the other hand counteract superordinate claims to knowledge such as the orientation towards the common good. If each individual point of view is regarded as equivalent knowledge, then e.g. politicians hardly work towards a collective good, since balancing these demands is no longer possible (Campbell 2012, p. 140), just as little as an opportunity for criticism.

References

Alexander ER (2001) What do planners need to know? J Plan Educ Res 20(3):376–380. https://doi.org/10.1177/0739456X0102000309

Alexander ER (2016) There is no planning—only planning practices: notes for spatial planning theories. Plann Theor 15(1):91–103. https://doi.org/10.1177/1473095215594617

Amin A, Roberts J (2008) Knowing in action: beyond communities of practice. Res Policy 37(2):353–369. https://doi.org/10.1016/j.respol.2007.11.003

Aristotle (1975) Doctrine of the inference or the first analytics [1921]. Felix Meiner, Hamburg

Bacon F (1990) Novum organum [1620]. Felix Meiner, Hamburg

Barker G, Kitcher P (2014) Philosophy of science—a New introduction. Oxford University Press, New York
Campbell H (2012) Planning to change the world: between knowledge and action lies synthesis. J Plann Educ Res 32(2):135–146. https://doi.org/10.1177/0739456X11436347
Chisholm R (1957) Perceiving: a philosophical study. Cornell University Press, Ithaca, N.Y.
Comesaña J, Klein P (2019) Scepticism. The Stanford Encyclopedia of Philosophy https://plato.stanford.edu/archives/win2019/entries/skepticism/. Accessed 22 Jan 2021
Chugh R, Wibowo S, Srimannarayana G (2015) Mandating the transfer of tacit knowledge in Australian universities. J Organ Knowl Manage. https://doi.org/10.5171/2015.297669
Dalton L (2007) Preparing planners for the breadth of practice. J Am Plann Assoc 73(1):35–48. https://doi.org/10.1080/01944360708976135
Davoudi S (2015) Planning as practice of knowing. Plann Theor 14(3):316–331. https://doi.org/10.1177/1473095215575919
Descartes R (1960) Von der Methode des richtigen Verstandesgebrauchs und der wissenschaftlichen Forschung [On the method of right use of reason and scientific research] [1637]. Felix Meiner, Hamburg
Descartes R (2009) Meditationen [Meditations] [1641]. Felix Meiner, Hamburg
Detel W (2014) Grundkurs Philosophie. Bd. 4, Erkenntnis- und Wissenschaftstheorie [Basic Course Philosophy, vol. 4, epistemology and philosophy of science]. Reclam, Stuttgart
Evans R, Marvin S (2006) Researching the sustainable city: three models of interdisciplinarity. Environ Plan A 38(6):1009–1028. https://doi.org/10.1068/a37317
Faludi A, Waterhout B (2006) Introducing evidence-based planning. disP 42(165):4–13. https://doi.org/10.1080/02513625.2006.10556950
Flyvbjerg B (1998) Rationality and power: democracy in practice. University of Chicago Press, Chicago
Forester J (1989) Planning in the face of power. University of California Press, Berkeley
Foucault M (1971) The order of things. An archaeology of the human sciences. Suhrkamp, Frankfurt a. M.
Foucault M (1977) Überwachen und Strafen—die Geburt des Gefängnisses [Discipline and punish-the birth of prison]. Suhrkamp, Frankfurt a. M
Gettier E (1963) Is justified true belief knowledge? Analysis 23:121–123
Guignon CB (1991) Pragmatism or hermeneutics? Epistemology after foundationalism. In: Hiley DR, Bohman J, Shusterman R (ed) The interpretive turn: philosophy, science, culture. Cornell University Press, New York
Innes J (2004) Consensus building: clarifications for the critics. Plann Theor 3(1):5–20. https://doi.org/10.1177/1473095204042315
Legg C, Hookway C (2020) Pragmatism. The Stanford Encyclopedia of Philosophy. https://plato.stanford.edu/archives/fall2020/entries/skepticism/. Accessed 22 Jan 2021
Lindblom CE, Cohen DK (1979) Usable knowledge: social science and social problem solving. Yale University Press, New Haven
Merton RK (1937) The sociology of knowledge. Isis 27(3):493–503. https://doi.org/10.1086/347276
Office of the deputy Prime Minister (ODPM) (2004) Planning policy statement 12: local development framework. ODPM, London
Polanyi M (1966) The tacit dimension. University of Chicago Press, Chicago/London
Popper K (1963) Conjectures and refutations. Random House, New York
Russell B (2009) History of western philosophy. Routledge, London
Rydin Y (2007) Re-examining the role of knowledge within planning theory. Plann Theor 6(1):52–68. https://doi.org/10.1177/1473095207075161
Sahakian W (1968) History of philosophy. Barnes & Noble, New York
Sandercock L (1998) Towards cosmopolis. Wiley, London
Schülein JA, Reitze S (2012) Wissenschaftstheorie für Einsteiger [Theory of Science for Beginners], fifth UTB, Vienna
Solomon R, Higgins K (1996) A short history of philosophy. Oxford University Press, Oxford

4 Value Orientation in Spatial Planning

Abstract

The field of action of spatial planning as an interface between future-oriented solutions to complex existing problems is the arena of ethical and moral orientations and alignments. However, planners cannot make these ethical decisions alone. Orientation towards a common good in particular requires further differentiation in this sense. The three positions of normative ethics for evaluating an action also differ significantly when it comes to planning: Virtue ethics, which primarily values intention, deontology, which addresses the action itself, and utilitarianism, which deals with its consequences.

Keywords

Values • Future-orientation • Planning ethics

Knowledge networks as deeply normative Knowledge is recognised by Davoudi (2015, p. 8) as something contextual of space and time. At the same time, it is deeply normative, even though there may be a causal relationship. When planning is understood as a practice of knowing, it implies that everyone in the process has knowledge and the boundaries of knowledge are rather vague. The totality of those who know results in socially constructed knowledge—that which is recognised as knowledge by society (Davoudi 2015, p. 8). From this and from the interaction of knowledge and action, there is the possibility of a practical assessment of projects, which e.g. are carried out by virtuous political actors. At the same time, Davoudi (2015, p. 9)—following Forester (2012)—sees knowledge as something pragmatic, in the sense of focusing on the consequences of the action instead of considering the intention of the action. However, it is always purposeful and object-oriented (Davoudi 2015, p. 9). This pragmatic and purposeful perspective on knowledge in planning means that reflection is an important dimension in practice. Forester (2012, p. 9) understands the following fact by reflection in action: "as planners, we are not only thinking when we act but also doing when we act" (Davoudi

M. Levin-Keitel and L. Behrend, *The Topology of Planning Theories*, The Urban Book Series, https://doi.org/10.1007/978-3-031-37857-7_4

2015, p. 9). Following Levi Strauss (1966, p. 66), Davoudi compares planners with bricoleurs. Strauss uses this as a metaphor to show that actors use different parts (different knowledge, instruments, material) to create a whole that is fit for purpose. However, the purpose results in part from the availability of the respective resources and the use of the resources only becomes visible in part during the process (Freeman 2007, p. 486). The relationship between power is not seen as a direct connection between the powerful and the powerless (Davoudi 2015, p. 10). Foucault (1977) already presented these complex playing fields on which the actors influence each other. Accordingly, knowledge and actions are also influenced and planning as a practice of knowledge "prescribes what is to be done and what is to be known" (Davoudi 2015, p. 10). Foucault has characterised this as the judiciary (the regulative aspects of an action) and the verdicative (the creation of a legitimising discourse) character of practice (Foucault 1980, p. 47). Power is not to be associated in a purely negative way—it can also play a productive role in relation to actions (Davoudi 2015, p. 11). Power is represented through various forms—such as language, signs, metaphors and symbols (Davoudi 2015, p. 11). These are all forms through which planners also express their knowledge and actions and play a role in the assessment (Davoudi 2015, p. 11). However, these are selective abstractions that cannot do justice to everyone. Furthermore, power is represented both by formal rules and the system as a whole, and by informal rules. Planners should, for instance, plan for the common good (whatever this might be) and take all concerns into account attentively, skilfully and comprehensively - but this does not mean that the result will be good. It is therefore more of a support for everyday life in order to carry out the processes as effectively as possible. Flyvbjerg (2001, p. 57) has already shown that something must already be perceived as good on the basis of certain values in order for something else to be described as good in comparison. Accordingly, the entire process is normatively guided from the outset.

4.1 The Transition from Knowledge to Action

From knowledge to action It is only through the synthesis of the present and the future that knowledge and action are related. Dewey (1954) already recognised the problem that the analysis always aims at the past and therefore comes too late for future planning. It can therefore only provide information, but cannot provide any basis for evaluation with regard to the future target state (Schön 1991). Campbell (2012, p. 141) tries to solve this problem by examining and comparing analysis and synthesis. The key difference is that analysis requires dividing reality in bits and pieces in order to examine them, while synthesis looks at reality by a complex understanding of the system as such. The world of analysis is therefore fragmented and draws narrow boundaries of competence, which leads to increasing specialization—which can also lead to a very narrow field of vision (Schön 1991, p. 60). The world of synthesis is broader, but it also risks being too broad and exceeding the limits of knowledge. "It follows from the preceding that while the results of analysis point to the complexity of the world, synthesis seeks clarity" (Campbell 2012,

p. 141). Another important distinction is that of the method of thinking. Analysis is based on explanation, synthesis on reasoning—this distinction of Campbell is also based on Dewey and supports the perspective that analysis looks to the past through explanation, while synthesis looks to the future through reasoning (Campbell 2012, p. 142). *Reasoning* involves the past but is geared towards normative rather than explanatory purposes. Kwame Anthony Appiah (2006, p. 73) summarises it as follows: "Reasoning—by which I mean the public act of changing stated justifications—comes in not when we are going in the usual way, but when we are thinking about change." The justification is the foundation on which the argument is built. It also follows that the analysis is much more based on a robust methodology, while the synthesis is also mainly based on judgment and ethics (Campbell 2012, p. 142). Accordingly, analytical knowledge is easier to describe than synthetic knowledge, which often remains vague, almost mysterious (Campbell 2012, p. 143; Schön 1991). However, analytical knowledge without synthetic knowledge is of little help, as otherwise only the description of the past remains—here only the argumentation for the future helps, whereby it is not about the one *good*, but about improving the current situation (Campbell 2012, p. 143). Recently, there has been an excessive focus on analytical knowledge (Campbell 2012, p. 143). This has led to a repression of the ability to reason and, above all, of the question of what needs to be done. "Analysis alone may produce knowledge. But analysis needs synthesis to become meaningful and to have effect and impact in the world" (Campbell 2012, p. 143). In accordance with the close connection between knowledge and action with values, they need to be explicitly considered in spatial planning.

Values in Action The field of action of spatial planning as an interface between future-oriented solutions to complex problems and political-democratic decision-making is the arena of ethical-moral orientations and alignments. This is further reinforced by the need to prepare decisions, since planners are required to set desirable goals and justify them. However, any goal setting is a deeply normative process because it is based on individual or collective value decisions. Despite this area of conflict, the value-laden nature of planning and its theoretical reflection in planning theories is currently little discussed. It remains open to what extent values appear in planning and, above all, when and which influences are compatible with the planning ethos. The reflection of normative ethical foundations of planning theories also plays a decisive role. This question ties directly to what planners can know at all, from which the question of what planners can do (at all) can be derived.

Practical Syllogism The question of what planners can know aims at the basis of their knowledge, while the question of what they can do raises the question of the goal. The naturalistic fallacy, i.e. the statement that no direct deduction about actions can be made from existing knowledge (no *what ought to be* follows from *what is*), shows that the question of the goal is completely different from the question of the knowledge base. The knowledge base serves only as a foundation for deciding on spatial planning goals. Accordingly, the question of the goal is a question of value. But how do planners achieve these goals? The practical syllogism (Nussbaum 1978), which goes back to Aristotle, could be helpful as an explanatory model. It represents a model of action theory and essentially depicts human action. Unlike models of logic, which make a statement by reasoning, the practical syllogism focuses on an action. The result is a teleological explanation; i.e. an explanation that refers to goals (Nussbaum 1978, p. 59 ff), whereby the connection between goal and means in an action is shown. In an example, this could be explained in spatial planning as follows:

Major premise: Densification is desirable.

Minor premise: Erecting buildings in backyards means post-densification.

Conclusion: Erecting buildings in backyards is desirable.

In the upper premise, the major premise, an intention is expressed, which represents a *want* or *ought to be* and thus implies a value concept. It is therefore a normative premise, since when asked why the respective goal should (or wants) be achieved, reference is made to desires or duties or moral imperatives (Corcilius 2008). In the lower premise, the minor premise, the knowledge of a means to achieve the respective goal is expressed. A requirement will be answered here by reference to experience. The sociologist Gerhard Schule therefore sees a *knowledge component*, a *will component* and an *action component.* The practical syllogism is clearly related to human action (Corcilius 2008) and is used in the human sciences, insofar as they are interested in human actions, and also in the technical sciences, insofar as they are goal-oriented (Hardie 1980). Along with the normative premise comes an assessment, which is that the goal in question is considered morally good. In addition, the goal in the cognitive premise is linked to a means of achieving the goal. In this way, the respective means are recognised as being suitable for achieving the goal. In addition, a variety of other means can fulfil this function. However, goals, means or functions are not observable and are based on the interpretation of facts, states and processes. Figure 4.1 shows how the practical syllogism works. The descriptive premise maps the current state of knowledge, which is converted into an action judgment (conclusion). The prescriptive premise represents a core component of the practical syllogism: Depending on what is assumed to be morally good or desirable (the moral position), the corresponding judgment about the action to be taken diverges, although the data or information basis is the same.

4.2 Moral Positions and Action-Orientation

Moral positions of normative ethics However, the practical syllogism does not give any instructions about what should be done, but offers a model for mapping the decision-making process (see Fig. 4.1). Rather, the decision is based on the determination of a target state as normatively desirable and the interpretation of

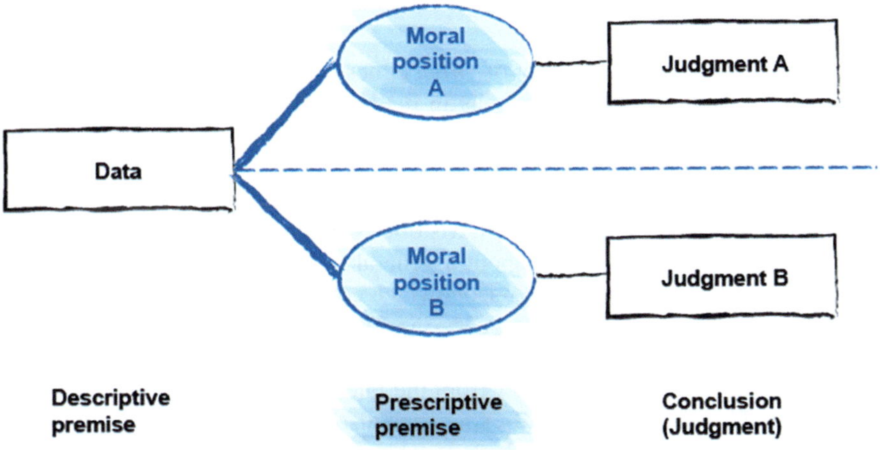

Fig. 4.1 The practical syllogism as a basic model of a value judgment. (Own illustration)

the respective means as suitable for achieving this target state. However, which moral positions of normative ethics are there for evaluating an action? And can perspectives be shown as to what planners should do? The three classic positions are virtue ethics, which primarily considers and evaluates the intention of an action, deontology, which focuses on the action itself, and utilitarianism, which deals with the consequences of an action (see Fig. 4.2). All three moral positions play an important role in spatial planning, but differ significantly in their reference points (moral positions of motivation, action and outcome).

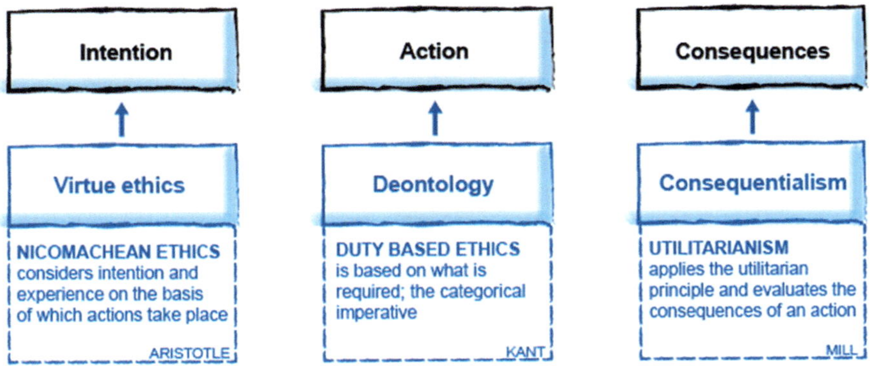

Fig. 4.2 Moral positions of normative ethics and their reference points. (Own illustration)

4.2.1 Virtue Ethics

Nicomachean Ethics The concept of virtue ethics is closely linked to Aristotle and goes back primarily to his discussions in his book Nicomachean Ethics (Russell 2009). This position of normative ethics reflects human action. Aristotle attempts to improve action through analysis and insight based on experience (Aristotle NE). It encourages people to achieve the most important goal in life, which for Aristotle is true happiness (eudaimonia) (Aristotle NE. I 5, 1097b20). The experience is significant insofar as it is not about a general principle, but always about concrete actions—the individual case—from which constants and regularities are to be worked out (Aristotle NE. I 7, 1098a25–35). Whether an action is correct cannot be seen from the success of the action, such as the attainment of wealth or political power (Russell 2009), but rather from the intention with which this action was carried out. This important role of experience also leads to criticism of an idea of the good in the Platonic sense, which is why Aristotle criticises Plato. For him, ethics is based on the analysis of human action and at the same time he associates ethics with a far less precise claim to knowledge than, for example, with the exact sciences (e.g. mathematics) (Aristotle NE). Accordingly, ethics does not prescribe any actions and cannot do so either, since not every detail of the individual case can be taken into account in general, but the individual conscience must make the final decision (Lutz-Bachmann 2013, p. 56). Rather, ethics becomes a compass, comparable to moral intention.

Significance of experience and voluntary action Now the question arises as to what the good mentioned by Aristotle is, and accordingly he undertakes a comprehensive analysis of human action himself. In his opinion human striving has a structure that is oriented towards a goal (Greek: *telos*) (Aristotle NE). This is based on the natural-philosophical assumption, which was widespread at the time that all life strives towards a goal corresponding to the respective species, including humans. Humans are characterised by language, reason, sensory perception, but also by the ability to remember and reflect (Aristotle NE). Humans are also characterised by the ability to choose between different goals; this is possible on the basis of practical knowledge in everyday life. In order to arrive at the *good*, Aristotle requires a further distinction. Namely between an action whose goal (*telos*) lies outside of the action and the doing, whereby the goal is already achieved with the performance of the action (Aristotle NE). The former means an activity in the technical sense that produces something (*poiesis*), while the latter aims at a type of action that can be understood in terms of *praxis*, in which man realises himself with his peculiar dispositions (Aristotle NE). Moral action now takes place in *praxis*, i.e. the broad concept of action and not in the narrower *poiesis*. The goal of moral practice lies in good action itself and thus in the success of self-realisation (Aristotle NE). The close integration of planning into the concepts of *poiesis* and *praxis* becomes immediately clear. Although planning is a process of the technical production of spatio-temporal structures, planning can primarily be assigned to the field of *praxis* and accordingly always has a normative character, since the goal of planning actions lies outside the action itself.

Ethical Virtues as a Goal of Action The fulfilled life—*Eudaimonia*—is the term that evaluates the action morally positive (Aristotle NE. I 5, 1097b, p. 5–30). However, the goal is not a final goal, but it is only achieved through an *actively* lived life (as Hannah Arendt also understands it in her writing The Human Condition (1958) as *vita activa*) and a constantly ongoing action. Life plans geared toward wealth, power, or pleasure do not meet the criterion because they are not action goals pursued for their own sake. They therefore lack the moral quality (Aristotle NE. I). Only bliss, *eudaimonia*, fulfils them. What this consists of, however, is quite vague. More specifically, Aristotle tries to capture them through the ethical virtues. Virtues are not natural dispositions but are acquired, i. e. they are dispositions that enable man to perform appropriate actions (Aristotle NE. I 13, 1102a5-1103a10) (a way of thinking that Davoudi (2015) has also adopted). For example the virtues of *prudence, bravery* or *justice* (Aristotle NE. II–V, 1103a14–1138b15). These are acquired through practice, imitation and insight and, together with cleverness, shape the respective character of the person (Lutz-Bachmann 2013, p. 61). This makes it possible that in each individual case the right action, the right goal and the means leading to the goal can be balanced (Aristotle NE. VI, 1138b16–1144a10). Now the question arises whether virtues of this kind are relevant for planners at all or whether the predispositions in the area of technical knowledge are decisive. This knowledge is certainly important, but its application arises from actions, which may be more or less virtuous. Accordingly, the combination of prudence and virtuous action is essential for planners, since the intention of actions and thus of projects that are aimed at the future can first and foremost be normatively justified and evaluated on the basis of their intention.

Criticism in a nutshell Criticism is primarily aimed at four aspects (Lutz-Bachmann 2013, p. 62 f.): First, Aristotle's approach appears circular, since the good is used both in the justification and in the explication of the good and in the sense of the criterion sought; second, the doctrine of virtue is in some way tied to the value preferences of the given society; third, there is no ethical claim to universality in the sense of moral correctness; fourth, the normative idea of moral obligation is missing.

Implications for planning theories The consideration of Aristotelian virtue ethics leads to some interesting aspects for the planning-theoretical consideration of planning value bases. On the one hand, it opens the debate about which normative-ethical perspective on actions in planning should be given preference (deontological and utilitarian perspectives are considered below). In planning theory, it is true that normative elements and the importance of ethics in planning processes are dealt with, or even justice is demanded, but there is often no in-depth clarification of the underlying assumptions. Demanding justice or moral action is meaningless unless the particular position of justice or morality is also presented (What is just? What idea of morality is held?). On the other hand, the consideration of virtue ethics and the associated focus on the intention of the action provides a very fruitful approach for the evaluation of planning action. Representation for the common good can also be attributed to virtue ethics, i.e. the intention to take the good of all into account. However, it seems difficult to determine and then evaluate the intentions of planners. Planners define the goals of their measures, from which the intentions can be derived. However, there is also a fundamental problem at this level, which consists in the impossibility of experiencing the actual intentions of the planners, which do not necessarily have to correspond to personal intentions. Nevertheless, the concept of virtue ethics offers some interesting aspects, since looking at an action from the intention creates new perspectives and the planner as an actor of the action comes to the forefront more than other conceptions do.

4.2.2 Deontology

Ethics of duty Deontology derives conceptually from the Greek *dei* and means something like: *one must or one should*—it is therefore about something obligatory (Lutz-Bachmann 2013, p. 76 f.). The corresponding ethics of duty is closely related to Immanuel Kant, whose theory will now only be outlined. *Good,* as a moral attribute, is used in Kant's ethics for the human will if it is only determined by insight into practical reason (Kant, GMS). A good will is characterised by the fact that it is free from any influences and its content „von der Einsicht der praktischen Vernunft in das ‚moralische Gesetz ' des Handelns bestimmt [ist]" ["[is] determined by the insight of practical reason into the 'moral law' of action"] (Lutz-Bachmann 2013, p. 91). The question now is what is to be understood by moral action. To this end, Kant deals with *moral duty* and thus equates morality with an imperative or commandment. He knows three ways in which a person can fulfil their moral duty in their actions (Kant GMS): First, there is the possibility of outwardly acting dutifully; here the motives can deviate accordingly from the action or do not have to follow from moral insight. Second, it can be an act of duty, and thus through insight into what is morally imperative, although other motives are also pursued. Third, an act may stem from duty alone; that is, on the basis of insight into what is morally required and only from this for the sake of duty, so to speak. For Kant only this third case is a moral action.

Significance of Reason For Kant, the practical reason of humans represents the highest principle of the validity of morality (Kant GMS). It recognises what is morally required and formulates this in the form of a law, whereby it acts as self-legislation and thus fulfils autonomy (Kant GMS). This autonomy of self-legislation is to be equated with the free will in us, which, however, only remains free by listening to morally legislative and autonomous practical reason. This is not to be equated with the famous categorical imperative, which can be seen more as a standard to check whether the respective action is morally correct or necessary (Lutz-Bachmann 2013, p. 93). The categorical imperative is formulated by Kant in different formulas. Each of these serves the above purpose; however, two are fundamental. 1. "Handle nur nach derjenigen Maxime, durch die du zugleich wollen kannst, dass sie ein allgemeines Gesetz werde." ["Only act according to that maxim which you can also want to become a general law"] (Kant GMS AA 04, p. 421) and 2. "Handle so, dass du die Menschheit sowohl in deiner Person, als in der Person eines jeden anderen jederzeit zugleich als Zweck, niemals bloß als Mittel brauchst. ["Act in such a way that you protect humanity both in your person and in the person of each other at the same time as an end, never just as a means"] (Kant, AA 04, p. 429). From this follows, for instance, the supreme moral principle, respect for human dignity and the freedom of the individual in the form of a legal principle. For spatial planning, the perspective and evaluation of the actual action is of particular interest. The advantage of this, compared to virtue ethics and utilitarianism, is the focus on the actual action. This circumvents the difficulty mentioned that, for instance, the actual intention can only be derived indirectly and is not directly recognisable. In the planning process, planners can focus on individual actions in the evaluation and morally justify themselves in this way. The first form of the categorical imperative, which can be used as a standard for the actions of planners, is of particular importance. In Kant's model, reason plays a particularly important role, which denotes the ability to draw conclusions and carry out self-examination, which in turn are independent of experience. For planners, reason is therefore a key aspect in their everyday work, since a large number of decisions (especially future-oriented ones) not only depend on experience, but also have to be made outside of experience—through reason.

Criticism in a nutshell Criticism was levelled in particular at the rigourism in Kant's formulation of acting solely for the sake of duty (Lutz-Bachmann 2013, p. 93). However, Kant's ethics has proven to be extraordinarily long-lived and expandable, and its principles are still treated as an important point of reference for many works.

Implications for planning theories The consideration and evaluation of an actual action is of great advantage for planning theories. As mentioned, this is much easier to do than dealing with the intentions or possible consequences. Accordingly, actions from practice can be presented and evaluated on a theoretical level. At the same time, this focus on the respective action leads to the difficulty associated with not considering consequences. Since planning also deals with future consequences and concerns, it may seem strange not to include these estimates in the evaluation of actions. Despite this, one would have to limit oneself to pure action if a deontological perspective is followed. In this respect, planners and academics are required to evaluate the actions from the point of view of the action and to ignore other considerations. For instance, planners should not lie in negotiations or try to otherwise influence their counterpart in their favour. This is already problematic insofar as planners not only passively mediate and try to balance different demands in planning processes, but also incorporate their own content and positions. In particular, the claim, which is based on technical knowledge, leads to an influence on other actors and a special position of power in the planning process. The close exchange with politicians should also be mentioned, which can be significantly influenced by planners due to the preparation of decisions. In principle, this is not a problem as long as planners do not take advantage of this position of power to put their own interests first. Accordingly, the model of deontology also shows a strong focus on the individual, who should act morally through the conscientious use of reason and the orientation of his actions along the categorical imperative.

4.2.3 Utilitarianism

Ethical consequentialism Utilitarianism deals with morality in an entirely different way than the two previous models. Instead of an upstream search for the morally good or the virtues, it looks for a rational criterion that claims universal validity and clearly determines what is right (Mill 2002 (1863]). This should be determined based on the consequences that result from the action—a distinction is made between the intended and the foreseeable consequences. The former are considered retrospectively, the latter prospectively. In addition, a distinction is made between closer and further consequences of action, whereby the latter cannot be taken into account (Mill 2002 [1863]). The pure consideration of the consequences of action leads to the fact that utilitarianism is counted among ethical consequentialism. The name utilitarianism can be traced back to the fact that the utility principle is the principle according to which the consequences of an action are evaluated—something is good if it benefits someone or is within the framework of the common interests of those affected by the action (Mill 2002 [1863]). However, this principle of utility can be applied differently to the consequences of action. Either with regard to maximum happiness in the sense of an accumulation of pleasure or with regard to the well-being of everyone.

Utility principle John Stuart Mill has massively differentiated the utility principle already formulated by Jeremy Bentham. He has developed four sub-principles that are necessary for the justification of the principle of morally correct action and must be linked (Mill 2002 [1863]): (1) The principle of consequentialism—the focus on the consequences of an action in order to evaluate it. (2) The principle of usefulness or utility—in order to be judged as correct, actions must have some specifiable utility. A distinction can be made between foreseeable and unforeseeable but ethically relevant consequences. (3) It is measured whether an action increases happiness or prevents unhappiness—that is, to what extent pleasure and joy are increased quantitatively and qualitatively and pain is avoided. (4) The principle of general well-being—not just considering the actor, but whether happiness/joy can be increased or unhappiness/suffering decreased for a large number of those affected by the action. In a sense, the goal is the greatest possible happiness for the greatest possible number. A perspective that is also reflected in the planning. However, his approach raises various problems, which are mainly based on the fact that moral-philosophical premises, anthropological assumptions about human nature and moral value judgments are linked (Lutz-Bachmann 2013, p. 69). Furthermore, normative statements about the increase in happiness/unhappiness are linked to it and an attempt is made to underpin this with empirical-descriptive factual expectations (Lutz-Bachmann 2013, p. 69). Planners take a similar approach. In doing so, they link their value judgments with assumptions about the way of life and everyday habits as well as movements in space. At the same time, descriptive material is used to further support the conclusions and make them suitable for a majority. This results in a mixture that can no longer be broken down into its individual components. Accordingly, it is extremely difficult to determine which aspects have how much influence on the (ethical) result of the planning.

Criticism in a nutshell The consequence of these many connections is a problem in justifying utilitarianism. Lutz-Bachmann (2013, p. 70 f.) deals primarily with four classic criticisms of this theory: (1) The initial thesis of utilitarianism is not sufficiently justified—it remains unclear to what extent the assessment of an action as useful can be used to conclude that it is binding. In addition, no normative knowledge can be derived from the assumptions about reality. (2) It remains unclear to what extent qualitative feelings such as pain, joy, sadness, etc. are transferred to a quantitative calculation, which is absolutely necessary for balancing different demands. (3) Mill does not explain how the relationship between the objectively presented interests of people and the actually articulated interests of individuals is to be thought of. Rather, it is a question of assumptions and, at best, statistical considerations. Verbalisation in actions is generally ignored. (4) It remains unclear how human rights and moral dignity are to be granted to every human being, regardless of the consequences of his or her actions. Later utilitarian responses to this critique were formulated, for instance, by action and rule utilitarianism. The former tries to look at each individual action and to analyse and then evaluate the consequences for those affected. However, a purely quantitative benefit calculation, as undertaken here, is fundamentally misguided when it comes to ethical questions of moral correctness (Lutz-Bachmann 2013, p. 73). The latter tries to show whether a general well-being follows from certain actions. Although this position is comparatively understandable, the reasons that lead to a positive assessment of the respective action remain ethically unclear, and it also remains unclear why a rule that has been tested in this way should be followed in a normatively binding manner (Lutz-Bachmann 2013, p. 74). Preference utilitarianism, as advocated by Peter Singer (1979) for instance, cannot provide an answer to this question either.

Implications for planning theories Utilitarianism is extremely interesting and problematic at the same time for planning theory. There is no question that planning is always directed towards the future and tries to balance various interests also in a normative and moral way. Accordingly, utilitarianism as the guiding principle for the evaluation of planning actions is obvious. However, this has a number of demonstrated theoretical difficulties, which also become visible when applied to planning action. So it cannot be explained to what extent the transfer of qualitative perceptions (e.g. aesthetics) can be transferred into a quantitative principle of balancing, or how the actually articulated interests of individuals (e.g. from participation processes) can be transferred into the abstract common good. At the same time, it is also unclear how future conditions should be compared and balanced if it is unclear whether and to what extent they will occur at all and what the consequences will be.

Summary of moral positions and their relationship to planning theories Which moral position does spatial planning take and how can this be further deepened? While virtue ethics tend to emphasise the general intention of planning, i.e. the orientation towards the common good as such, deontology and utilitarianism are more based on individual moral positions, such as the fulfilment of duties or the principle of utility. Spatial planning tries, similar to utilitarianism, to differentiate between statements about the intended and foreseeable consequences of actions and to balance them, which then entails a decision regarding the greatest possible common good. As can be seen in the case of utilitarianism, this is an elusive position in theory and even more so in practice. Too many imponderables remain unclear and, in particular, when considering the planning process, it is not clear which criteria should be used and how the mixture of qualitative and quantitative concerns should be dealt with. The mixing of the three positions of normative ethics in planning processes becomes interesting. Planning can be based on a specific intention, an attempt can be made to compare the consequences of planning measures and the action itself can be considered and evaluated (e.g. through the choice of means). In the end, it remains ambiguous how planning should ultimately evaluate, even if the importance of ethics for planning is undisputed, since it is important to present a moral rationale for decisions in order to legitimise them (Beatley 1989, p. 2). Possibly, it is not only necessary to consider a single position, but to consider and evaluate all three positions. This results in an enormous normative network, which could explain why so few planning theorists have dared to unravel. This fact is also criticised in more recent publications: “Regardless of these differently motivated and modelled approaches, the research and publication situation is anything but satisfactory” (Berr 2017, p. 2). This fact becomes even more complex if we include the individual level of the value decisions of individual planners.

4.3 Between Planning Knowledge and Democratic Action

Knowledge and democratic decisions Today’s decisions, especially from a spatial planning point of view, shape and enable distinct future possibilities—of course not in a deterministic way and without claim to exact prediction. Nevertheless, the built environment that is realised here and now, the rules and regulations of space and its uses, the historical significance and cultural symbolism determine future development to a non-inconsiderable extent. However, on which basis are decisions made in a democratic system? As has been shown so far, knowledge in a variety of forms and the underlying values play a decisive role. Not only the transition from knowledge to action is value-laden, knowledge itself is already based on value decisions (Barker and Kitcher 2014). There is no such thing as value-free knowledge, since the very decisions about what counts as knowledge, what questions are considered of general interest and worth investigating, and the choice of methods and tools all contain values (Barker and Kitcher 2014). This also means that the basic assumptions in planning processes represent values. However, since planning decisions are social issues, there is a need for particular transparency and openness as to the values on which these decisions are based. Planning wrestles with the question of what proper decision-making should look like. In conversations with planners who are active in planning, the area of conflict becomes clear: Trained planners know very well that inner-city densification is preferable to building in the countryside, that building in flood-prone areas is not a sensible idea from a technical-rational point of view, and yet people want a house in the country with large shopping centres nearby or with unobstructed view of the seemingly peaceful river. This at least sometimes gives the impression that planners know about facts, but these are not enforceable in the majority society—colloquially the difference between (a planning) truth or the (often obligatory) participation in democratic decisions. In planning theory, this political dimension of decision-making has been the subject of many discussions for decades (e.g. 1998; Forester 1982).

Democratic dilemma of planning The relationship between spatial planning in the democratic dilemma between technical expertise and democratic opinion-forming as the basis for social decision-making is of crucial interest for all planning activities. The way in which people in regions and cities will act in the future and how these decisions will generally be made raises and reflects the core question of planning principles and their rationalities as well as paradigmatic assumptions (Behrend and Levin-Keitel 2020): Can't this unresolved dilemma be seen as a fundamental difference in planning-theoretical approaches? Are planning-related decisions based on their own claims to reality in the sense of their own rationality? Isn't this part of the rationalistic approach to planning? Or are planning decisions in all participatory and collaborative approaches not inherently political negotiations and persuasion that rely on majority-capable interventions? So doesn't it lead back to what is discussed under communicative or collaborative planning? The question of whether politics and in particular democratic decision-making follows technical requirements or democratic logic and the relationship between the two has not yet been sufficiently considered in planning sciences. Should decisions be made based on professional and technical knowledge, as some people know better? Or, is the idea of public interest and the perceptions of a majority of people the legitimate basis for decisions? These questions are based on different understandings of the state and ideas about the role of the state—in this case the role of spatial planning as a public administration, which represents a state authority.

Planning knowledge as technical expertise John Friedmann (1987, p. 10) dealt with questions of this type: „Are reason and democracy compatible? Can ordinary people be trusted to use their heads in the conduct of their own affairs, or is superior wisdom needed?" Planning implies knowledge about future states, or at least assumptions about future states. These are based e.g. on analyses, experiences or forecasts. As the Covid pandemic impressively demonstrates, these fundamentals are easily overwhelmed by unplanned events. Conversely, the scientific facts serve as argumentation and justification for planning decisions. Or, as Allmendinger (2017, p. 67) states: "The irrational forces of life—politics in particular—needed to be tamed through rationality and the rules of science. Planning was part of the means by which this could be achieved." The aim is to enable people to take control of their own environment through comprehensible and plausible decisions. The core of this consideration, and this is where the relationship to technical expertise arises, is the separation of content and process. Planners are not politically legitimate actors and derive their legitimacy from the technical competence and objectivity in the planning process (Faludi 1987; Meyerson and Banfield 1955). Planning should therefore be immune to power interests. This position, which planners develop, is derived from knowledge as the basis for rational decision-making processes. It is based on a rationalist perspective, which involves belief in objective knowledge and a reality that can be grasped: The disposal of knowledge, the production of knowledge and the possibility of objective knowledge through the collection of available information and facts. This perspective is in the tradition of positivism. Here it is considered necessary to recognise findings, which in turn should count as knowledge, only on the basis of "positive" findings and interpretations as such. They must therefore be traceable to sensory perceptible and verifiable findings.

Rationalistic planning is based on the claim that planning should meet rational, factual, objective and scientific criteria. Furthermore, planning should free people from subjective and emotional structures and help to control their environment by means of comprehensible and plausible decisions. The basis for this is the belief in the possibility of creating a rational order or solution for upcoming problems and that planning can contribute to the perfection of the world. These assumptions are already based on values, such as the objective that planning should meet rational and objective criteria. At the same time, however, it is completely unclear what is actually meant by the respective terms. For instance, how should a perfection of the world be imagined? Everyone has a different idea and different preferences as to how one should ideally live. In addition to these target states, whatever their nature, the selection of the means to achieve the target state is also deeply normatively charged. Contrary to what is assumed in rationalist planning, these do not result objectively from the desired target states, but are of a normative nature, just like the decisions about the targets. Thus, not only the respective decisions are normatively charged, but also the basic assumptions (such as the assumption that it is even possible to meet factual and objective criteria).

Planning knowledge in democratic decision-making Communicative planning theory has enjoyed particular popularity in the western world since the late 1990s and early 2000s (Mattila 2020, p. 1). The fundamental change in the understanding of planning is the introduction of other interest groups, other points of view, a different perspective than that of planning in the narrow sense of being a public institution. The special importance of consensus in this theory plays an essential role, as it is met with wide approval in the field of planning. Former and current proponents of the theory such as Forester (1989, 1993, 1999), Sager (1994, 2013, 1992), Innes and Booher (1995, 1999, 2010) or Healey (1996, 1997) agree that the different forms are based on the common idea that planning becomes more democratic, legitimate and fair through improved communication between the actors (e.g. planners in city councils, politicians, citizens, etc.) (Mattila 2020, p. 2). Forester (1993, p. ix) summarised this rethinking in planning as a strategic conflict between the various interest groups towards planning that creates consensus through communication in the sentence: "making sense together in a politically complex world."

Communicative planning takes the discourse theory developed by Jürgen Habermas as a starting point. Accordingly, one of the basic assumptions is that influential information is that which has been socially constructed by the community that uses it. Also, planning lacks standards to assess ordinary people's knowledge. Planners no longer appear as guiding stars of the planning process, but also as participants (with their expert knowledge) or as managers of the process. At its core, communicative planning assumes that planning should be made more democratic, more legitimate and more just by improving communication between the actors. All of this is also deeply charged with values. For example, it is completely unclear what is to be understood by justice—distributive justice, performance justice, etc.? Likewise, the determination of a democratic basic order as desirable is also a value-based decision. It becomes clear that even this general assumption from a Western perspective is accepted as valid only because of the social agreement that democracy is the desirable basic order (which seems like a highly contested claim at times). As with rationalist planning, it becomes clear that the basic assumptions are preceded by normative assumptions, which are then regarded as fixed. Ludwig Wittgenstein gives an understanding of this mode of action with his argumentative law of door and hinge. "*D. h. die Fragen, die wir stellen, und unsere Zweifel beruhen darauf, daß gewisse Sätze vom Zweifel ausgenommen sind, gleichsam die Angeln, in welchen jene sich bewegen*" [*"The questions that we raise and our doubts depend upon the fact that some propositions are exempt from doubt, are as it were like hinges on which those turn.*" (Wittgenstein 1989, p. 186). In Wittgenstein's case, this quote refers to scientific investigations, but its meaning can be transferred to the moral questions presented here. Because here, too, certain assumptions are made and not further questioned, since they are considered morally good or desirable. Based on these, interpretations and assignments are then made in the respective planning-theoretical forms, which in turn influence the means, instruments and other decisions (see Table 4.1).

Implications for planning theories In both cases, a rethinking of the ethical-normative orientation of planning is necessary for the acceptance of planning questions: Either to justify planning decisions or as part of the decision-making process. Given the daunting challenges humanity faces from the impacts of climate change and the need to adapt the way we live, work, use energy or move,

Table 4.1 Values in planning

Role of values	Examples
Determination of target states of planning	• Questions of use (What should be planned? Who to plan for?) • Decision about the inclusion of certain issues
Choosing the means to achieve the goals	• Choice of methods of collecting information • Decision on certain actions
Decision on how to deal with risks and uncertainties	• How much risk is accepted when forecasting future developments? • What counts as reliable knowledge and what doesn't?
Decision on how the project and the results of planning are to be communicated	• Selection of communication channels and forms of participation • Decisions about who should be involved

all spatially bound activities, the democratic dilemma becomes even clearer: Radical changes are needed, this seems to be knowledge that both society and science have agreed upon (considered as a doctrine). Nevertheless, measures and means to react to the consequences of climate change remain the subject of political debates, within political parties, in social media or other discourses. In both ways of decision-making, on the basis of technical expertise as well as democratic decision-making processes, planning is obliged to present its normative bases. In a way, planning could then act as an ethical compass (alongside technical knowledge) used to justify decisions. However, this requires a massive expansion of research into normative knowledge in planning and from planners. This is done on the basis of a wide variety of scales and is spatially bound. Truth alone and majority alone are not able to form a basis for planning that takes into account the democratic, future and pluralistic demands of society. Rather, planning is always to be seen in the field of conflict between the claim to the most comprehensive and reliable facts possible, with simultaneous reflection of this knowledge and the claim to the inclusion of the social and democratic will.

Consequences of the claim to truth How should decisions be made about the future of a society that moves between truth rationality and majority decisions? Certainly, no clear answer can be given to this question—after all, it is a question of a dilemma and a dilemma is characterised by its insoluble nature. A look at the consequences shows: If one follows the rational claim to truth, then spatial planning must establish and justify its approach. On the one hand, this includes why the rationality of planning (in the sense of its truth) should be so much wiser than other insights, rationalities and truths. An attempt is being made to attribute greater importance to planning knowledge due to technical knowledge. However, this technocratic view points to the philosophy of science perspective of realism, as it assumes that planners actually have insight into objective knowledge, a simultaneously extremely positivistic way of thinking. On the other hand, planning must explain what exactly this rationality is based on, under which objectives planning weighs up different demands, remodels a city, spatially organises a region, especially if it wants to develop an objective claim. If no objective truth can be reached, then the plan must at least show how a better approximation could be reached. It is therefore a question of more precise instruments that enable a significant improvement in the forecasts, etc. The aim must therefore be for planners to prepare and make decisions on the broadest possible basis of information; and that they also make transparent the values on which the balancing of the various concerns is based. Basically, however, the idea that planners can objectively recognise the right thing, the truth, based on their specialist knowledge does not seem tenable in a scientific understanding of the twenty-first century. More than an approximation is not conceivable and undermines a large number of planning activities.

Consequences of a majority claim After majority decisions have been made, planning must also deal with other challenges. On the one hand, when dealing with conflicts of interest, planning should take a standpoint as to why the rationality of the planners is not so important and other rationalities and perspectives on problems are just as important. In line with this need for explanation, planning must make the underlying goals, basic norms and values transparent and clarify in general how planning is positioned accordingly. On the other hand, it remains unclear how the planning is to achieve majorities among the actors. Is the role of planning then thrown back to lobbying to support one's own perspective on planning processes? Supported by court decisions that legitimise the reasoning behind the plan? Accordingly, planners only have one rationality among many others and need a normative justification. Otherwise it cannot be explained why this rationality is particularly important and it becomes difficult to convince other actors of this rationality. The question arises as to how this content can reach majorities and whether planners then have to act primarily as lobbyists for their point of view, a completely different self-image than *the helmsman* or the *moderating role.*

Values in planning Different roles of values can be abstracted from the previous train of thought. In Table 4.1 these are shown. Values are not only important when determining target states (i.e. the question of what should be planned at all and which concerns are to be included), but also when selecting means and instruments to achieve the defined goals and as a result the measures to be taken. However, the values do not only remain on the levels of goals and means. They are also relevant in the evaluation and the subsequent handling of risks and uncertainties, which are necessary to be found in a planning process. Predictions about future conditions always carry the risk of being inaccurate and accordingly the question arises as to what degree of uncertainty is tolerated. This also relates to what is recognised as reliable knowledge and what is not. Values are also important in communication. Especially when it is decided how a project and the results of the planning should be communicated. To do this, a selection of communication channels and suitable participation processes or forms of participation must be made, as well as who is involved and who is not. This depiction of different values makes it clear what role they play in planning, how diverse they are (fundamental decision of goals compared to decisions about communication that appear to be of secondary importance) and how extensively values appear in planning and thus have an influence everywhere. Every decision, no matter how small, in the entire process about actions, goals or tools is significantly influenced by values. These are located both at the individual level of the planners and are institutionally anchored, for example through legal bases.

Everything is normative All actions, not only in planning, are normative actions. This becomes particularly obvious in spatial planning, since various requirements and ideas are always compared here and a decision is then made about an action. Decisions are always based on value judgments. This is completely independent of which theoretical concept is considered the starting point. Even the attempt of rationalist planning to proceed as objectively and matter-of-fact as possible and to legitimise the decisions in planning processes in this way cannot distract from this. Every fundamental positing that is subsequently accepted is reconsidered in the normative decision in order to accept it as such. As has been made clear, while planning is shrouded in democratic dilemma, looking at the consequences offers a very clear picture of what needs to be done: An in-depth examination of the ethical orientation of planning. Instead of constantly trying to weigh up and somehow find the common good, planning should focus more on its own positioning. This can only be achieved through a well-founded professional training of the planners, which must be accompanied by the ability to think independently and critically. Planners must position themselves actively and more strongly than before in order to be able to meet future challenges such as climate change with all its direct and indirect consequences. Planners must not get carried away in the conflict between the pressure of the majority and their own hubris (of themselves or of politicians), but must represent their positions emphatically, but also give strong reasons for them. While climate-neutral goals tend to be discussed from a less spatial perspective (CO_2 reduction in a society, etc.), planning is forced to define places where a transformation takes place. This step from a general theoretical approach to a site-specific and argumentative justification is more like theorisation than formation in the field of planning. For this, planners need less technical and increasingly specialised expertise, but spirit in the sense of the ability to think spatially beyond their own area and situation.

References

Allmendinger P (2017) Planning theory. Palgrave, London

Appiah KA (2006) Cosmopolitanism: ethics in a world of strangers. Lane, London

Arendt H (1958) The human condition. The University of Chicago Press, Chicago

Aristotle (1985) Nicomachean ethics. Felix Meiner, Hamburg

Barker G, Kitcher P (2014) Philosophy of science—a new introduction. Oxford University Press, New York

Beatley T (1989) Environmental ethics and planning theory. 7 Plan Lit 4(1):1–32. https://doi.org/https://doi.org/10.1177/088541228900400101

Behrend L, Levin-Keitel M (2020) Planning as scientific discipline? Digging deep toward the bottom line of the debate. Plann Theor 19(3):306–323. https://doi.org/10.1177/1473095219897283
Berr K (2017) Architectural and planning ethics. Approaches, perspective points of view. Springer VS, Wiesbaden
Campbell H (2012) Planning to change the world: between knowledge and action lies synthesis. J Plan Educ Res 32(2):135–146. https://doi.org/10.1177/0739456X11436347
Corcilius K (2008) Two jobs for aristotle's practical syllogism? Hist Philos LogAl Anal 11(1):163–184. https://doi.org/10.30965/26664275-01101010
Davoudi S (2015) Planning as practice of knowing. Plann Theor 14(3):316–331. https://doi.org/10.1177/1473095215575919
Dewey J (1954) The public and its problems. Swallow Press/Ohio University Press, Athens; Faludi A (1987) A decision-centred view of environmental planning. Pergamon Press, Oxford; Flyvbjerg B (1998) Rationality and power: democracy in practice. University of Chicago Press, Chicago
Flyvbjerg B (2001) Making social science matter. Cambridge University Press, Cambridge; Forester J (1982) Planning in the face of power. J Am Plann Assoc 48(1):67–80. https://doi.org/10.1080/01944368208976167
Forester J (1989) Planning in the face of power. University of California Press, Berkeley
Forester J (1993) Critical theory, public policy and planning practice. State University of New York Press, New York
Forester J (1999) The deliberative practitioner: encouraging participatory planning progresses. MIT Press, Cambridge
Forester J (2012) On the theory and practice of critical pragmatism: deliberative practice and creative negotiations. Plann Theor 12(1):5–22. https://doi.org/10.1177/1473095212448750; Foucault M (1977) Discipline and Punish—The Birth of Prison. Suhrkamp, Frankfurt a. M.
Foucault M (1980) l´ Impossible prison. In: Perrot M (ed) Manuscript of a discussion round, 20 May 1978. Seuil, Paris
Freeman R (2007) Epistemological bricolage: how practitioners make sense of learning. Adm Soc 39(4):476–496. https://doi.org/10.1177/0095399707301857
Friedmann J (1987) Planning in the public domain: from knowledge to action. Princeton University Press, Princeton
Hardie WFR (1980) Aristotle´s ethical theory. Oxford University Press, Oxford
Healey P (1996) The communicative turn in planning theory and its implications for spatial strategy formation. Environ Plann B Plann Des 23(2):217–234. https://doi.org/10.1068/b230217
Healey P (1997) Collaborative planning. Shaping places in fragmented societies. Macmillan, New York
Innes JE (1995) Planning theory´s emerging paradigm: communicative action and interactive practice. J Plann Educ Res 14(3):183–189. https://doi.org/10.1177/0739456X9501400307
Innes JE, Booher DE (1999) Consensus building and complex adaptive systems. J Am Plann Assoc 65(4):412–423. https://doi.org/10.1080/01944369908976071
Innes JE, Booher DE (2010) Planning with complexity: an introduction to collaborative rationality for public policy. Routledge, London
Kant I: Kant's gesammelte Schriften. Hrsg. von der königlich preussischen Akademie der Wissenschaften [Kant's Collected Writings. Published by the Royal Prussian Academy of Sciences] [= AA]. G. Reimer, Berlin, 1900 ff.
Lutz-Bachmann M (2013) Grundkurs Philosophie. Bd. 7: Ethik [Basic course in philosophy. Vol 7: Ethics]. Reclam, Stuttgart
Mattila H (2020) Habermas revisited: resurrecting the contested roots of communicative planning theory. Prog Plann 141(1). https://doi.org/10.1016/j.progress.2019.04.001
Meyerson M, Banfield EC (1955) Politics, planning and the public interest—the case of public housing in Chicago. Free Press, New York
Nussbaum M (1978) Aristotle's De Motu animalium. Princeton University Press, Princeton
Russell B (2009) History of western philosophy. Routledge, London

Sager T (1992) Why plan? A multi-rationality foundation for planning. Scandinavian Housing and Plann Res 9(3):129–147. https://doi.org/10.1080/02815739208730300
Sager T (1994) Communicative planning theory. Ashgate, Farnham
Sager T (2013) Reviving critical planning theory. Routledge, London
Schön DA (1991) The reflective practitioner: how professionals think in action. Arena (Ashgate), Aldershot
Singer P (1979) Practical ethics. Cambridge University Press, Cambridge
Strauss AL (1966) Continual permutation of action. Transaction publishers, Piscataway
Wittgenstein L (1989) Über Gewißheit—Werkausgabe Band 8 [On Certainty—Edition Volume 8]. Suhrkamp, Frankfurt am Main

5 Planning Theories and Existing Systematisation Approaches

Abstract

Theories reflect different bodies of knowledge and are based on characteristic assumptions and basic values. Previous systematisation approaches place significant importance on the historical genesis and changes in planning understandings as well as in the distinction between substantive and procedural theories. Existing systematisations provide an overview of the different theoretical approaches over time or in practical research work. This does not yet correspond to a systematisation of planning knowledge with regard to scientific-theoretical foundations.

Keywords

Substantive planning theories • Procedural planning theories • Systematisation of planning theories

Theories in planning sciences A look at existing definitions of planning theory reveals the imprecision of the attribution of meaning to the terms: In most publications, planning theory is not clearly defined, but the term is set as if there was no need to specify its meaning. Allmendinger (2002) for instance starts directly with the definition of a Typology of Planning Theories, leaving the reader alone in imagining what exactly planning theories could be. Alexander (2010) directly addresses the question how planning theory affects practice and demonstrates different ways how knowledge is transferred into practice—implicitly arguing that planning theories represent this knowledge. There are manifold examples, especially books in planning sciences focus mainly on planning theories by addressing the term theory in an undefined way. Allmendinger (2009) for instance first asks what a theory can be and why theories are needed. The specification of what a planning theory could then be is answered with the help of a Typology of Planning Theories (*Exogenous Theory, Framing Theory, Social Theory*, *Social Scientific Philosophical*

M. Levin-Keitel and L. Behrend, *The Topology of Planning Theories*, The Urban Book Series, https://doi.org/10.1007/978-3-031-37857-7_5

Understandings and Indigenous Planning Theory), which, however, leaves the planning reference of a planning theory and its qualitative structure unanswered. These definitions do not provide an explanation of what characterises planning theories in particular and what their basic structure and content are. Merely Schönwandt (2008, p. 57ff) explains that theories show basic constructs of planning knowledge (concepts, statements and contexts) and these are defined within the discipline, an approach that still needs to be pursued. However, in most planning science publications there is no disciplinary discussion of the term planning theories. It becomes clear that the term planning theory is hardly adequately defined in planning sciences (Alexander 2016, p. 94), although the term is used, its implicit assignments of meaning are not made transparent. This also explains the relatively large range of implicit theoretical concepts and the diverse criticism of existing theoretical concepts. In the English-speaking world in particular, criticism of planning theories is based on their relationship to planning practice. This ranges from saying that planning theories are confusing, unhelpful, and inadequate (Neufville 1983, p. 35; Scott and Roweis 1977, p. 1097; Yiftachel 1989, p. 23; Lord 2014, p. 27) to assuming that theories contradict practical realities (Lord 2014, p. 30; Allmendinger and Tewdwr 1997, p. 802; Alexander 1997, p. 5). In conclusion, one can state that planning theories are meaningless for planning practice, since they only question planning practice (Gualini 2010). From the point of view of knowledge systematisation, this criticism, which is based on the relationship between planning theories and planning practice, is similar to the phenomenon of the naturalistic fallacy—if theory does not reliably determine *what should be* in the future, then it is confusing, unhelpful or meaningless.

5.1 Theories as Systematisation of Knowledge—Also in Planning?

Discussion about planning theories In planning sciences, three different approaches to the question of what planning theories are can be derived at first glance (Allmendinger 2009, p. 31; Hemmens 1980, p. 259 f.; Davoudi and Pendlebury 2010, p. 626; Friedmann 2008, p. 248 f.):

1. On the one hand, there are publications that state that there are no planning theories, only planning practices or practices of knowing. The thesis, widespread in planning sciences, that theories and practice represent two separate realities, and in particular that there can be a practice (or practices) without a theoretical perspective, is not tenable on the basis of a scientific understanding of critical realism. Practices are also based on theoretical approaches, both of which are more like two sides of the same coin (see Chaps. 1 and 2). Accordingly, it does not have to be either theory or practice, but the question that arises is that of the theoretical foundations. There is no neutral and objective view on practice in general, it is all theory-based based and influenced by different types of knowledge (based on experience, knowledge transfer, or the

like) and its application through practices: application of procedures, such as incremental steps (procedural theories), application of ideal–typical settlement structures, such as the garden city movement (substantive theories), application of market logics, such as compensation and replacement payments (social theories)? The basic assumption that there are no practices without theoretical approaches at least implies that there must be at least theoretical perspectives and approaches—even if there are no specifically differentiated planning theories.

2. Other publications support the thesis that there are planning theories, but that they have only been transferred to planning from other disciplines and are therefore, strictly speaking, not planning theories. This reflects the problem of a poorly defined epistemological object (see Chap. 1). Of course, the definition of what spatial planning means is the basis for classifying planning theories. The expression of transferring a theory also inevitably leads to the question of what this transfer implies, what constitutes the field into which it is transferred and which components play a role (e.g. social developments or a course of action). The resulting question is not whether there are any planning theories at all, but rather what is specifically related to planning. There is a lack of detailed discussion about the concept of planning behind (in a narrow or broad sense), which planning knowledge this theory covers (in terms of the actual or desired state) or which ethical principles the theory is based on.
3. Furthermore, other publications declare that there are innumerable, very different planning theories, with any knowledge being declared as planning theory, without further specification. Here, the terms knowledge and theories are used synonymously, a definition that is not very helpful from an epistemological perspective. While knowledge includes facts, theories and rules much more broadly (see Chap. 2), theories are only a part of it and designate a system of scientifically based statements that serve to explain sections of reality and the underlying laws (Schönwandt 2008, p. 62ff.).

The three approaches reveal fundamental inadequacies of planning theories, to the point of denying such constructs as planning theories. A look at the planning-theoretical debates in the relevant planning journals shows that currently individual approaches and tendencies usually located in planning or planning theories are used to differentiate oneself from them. So e.g. Rydin (2007, p. 58) describes the perspective of contemporary planning sciences more as an "opening-up" of different approaches than a "closing-down" of planning-specific claims. However, how the concept of theory itself needs to be enhanced is only marginally discussed (Alterman 2017). In summary, it can be stated that the deficiencies in the ontological, epistemological and ethical classification of what is traded as planning theories lead to naming either nothing or everything as planning theories.

Theories in science in general Building on the fundamental unity of theory and practice, however, no dedicated planning theories can yet be identified. In sciences in general, theories serve as a core component of disciplinary knowledge (Barker and Kitcher 2014). Theories are generally understood to be logical systems that do not necessarily represent absolute truth. They prove our current state of knowledge and can be invalidated by new evidence (Barker and Kitcher 2014). In scientific discourses it is necessary to define the central terms used—and to do so in as much detail as necessary. The aim of science is to create evidence and have it confirmed by a collective authority (e.g. the planner community) (Brehmer 2003, p. 686). This means that scientific findings are also based on the belief and trust ("faith") of the professional community, combined with the certainty that work has been done properly and that defined standards and methods are adhered to, i.e. a very clear normative setting (Brehmer 2003, p. 686). Strohschneider (2017), the former President of the German Research Foundation, put it this way: "The real problem is not that others hold other things to be true than we do (what else should the pluralism of modernity be?) … Modern research is pluralistic. It does not generate certainties, but methodically reliable knowledge" (Strohschneider 2017, p. 3 f.). That is, science does not speak of facts per se, but rather of ideas about facts, and that these ideas about facts correspond to theoretical constructs (i.e. theories). Or as Bunge aptly put it: "In science and technology, hypotheses and theories are confronted not with facts, but with empirical data relevant to the former conjoined with indicator hypotheses" (Bunge 2003, p. 60). Regarding facts as empirical data relevant to the respective theory again underlines the close unity of theory and practice. Building on this understanding of knowledge, theory describes a system of scientifically based statements that serve to explain sections of reality and the underlying laws and to make forecasts about the future, i.e. to justify these defined standards and show causalities (Wiesen 2003, p. 645). In general, theories consist of terms that are related to each other and together describe a context (Bunge et al. 1983, p. 44, transferred from Schönwandt (2002) to planning). If this context is self-contained, one speaks of a theory (in the narrower sense).

Planning theories as relevant basic constructs Planning theories as the organisation and systematisation of knowledge in the planning sciences are of particular importance, since they are understood with their contexts, relations and terms as the knowledge store of the discipline (Birch 2001; Davoudi 2015; Hopkins 2001; Pinson 2004). Nevertheless, it seems to be at least permanently in question what planning theories are, what they are used for and whether they have a dichotomous connection to planning practice. If one discusses planning theories as relevant basic constructs of a discipline, the knowledge store with the different forms of knowledge and their ontological and ethical characteristics can be further described for planning sciences. The hitherto inadequate definition of these knowledge elements in relation to planning theories has a negative impact on the recognition of planning as a science. A first necessary step in dealing with these basic constructs in planning sciences is to contribute to the way in which planning is carried out, what is planned and under which values and moral concepts these decisions are made *(gatekeeping)*. In the previous chapters, these scientific-theoretical foundations were explained explicitly. Furthermore, these basics serve as a key to a systematisation and enhancement of planning knowledge—the Topology of Planning Theories. In addition, a transparent debate about the genesis of theoretical approaches, their criteria and sources is necessary (*gate reporting*). In the scientific sense, the struggle for elementary standards is dependent on discursive debate and criticism and must therefore always be able to be further developed or discarded. The strong separation of theory and practice, i.e. the claim of action-orientation towards theories or the rejection of theories in general, turns out to be an obstacle and a complex challenge. Planning theories and their inherent knowledge of the current state are able to analyse and question planning practices; future-oriented knowledge, on the other hand, guides action on an ethical and moral basis and is of significant value for practice and how it is seen and interpreted. However, before we go into how these theoretical standards and the systematisation of planning knowledge can be developed and enhanced, the social added value of more well-founded planning sciences must be addressed. Finally, the systematisation of the planning knowledge base is not to be seen as a purely academic pull-up or the taxonomic urge of scientists: it rather requires the systematisation of planning knowledge for a qualified enhancement of spatial planning in the twenty-first century. Measured against the current challenges of our society, such as climate change, demographic change, sustainable land use or socially legitimate decision-making processes, spatial planning and science must also develop further, overcome previous paradigms and reinterpret premises such as the common good. Since the eighteenth century, it has been a widely held view that the application of knowledge based on scientific knowledge can contribute to the common good (Friedmann 1987, p. 51). Based on this former intention of Bentham, which also found its way into the scientific planning debate via Saint-Simon and is implemented (Friedmann 1987, p. 51 f.), the question can be asked from a scientific-historical perspective in the sense of Kuhn (1962), which paradigms currently dominate planning theory and what social relevance this question has. For planning, this means a systematisation regarding their theoretical conception. As an important actor with a high degree of social relevance, due to the preparatory function in far-reaching political decisions, an approach to their systematisation of specific knowledge must be made (Resnik 2000, p. 258). This appears to be particularly necessary because a joint theoretical conception is missing and a systematisation of the specific knowledge of planning is accordingly required in order to do justice to its claim as an application-oriented science serving society. Or as Alexander states: "What they perceive (quite rightly) as the theory–practice gap is the absence of systematic knowledge that translates the relatively general and abstract concepts and explanations of normative and positive theory into case specific prescriptions for action" (Alexander 1997, p. 5).

5.2 Systematisation of Planning Theories—And Their Relation to Planning Knowledge

Systematisations of planning knowledge Systematisations that classify, relate or group planning theories start from different perspectives, with specific goals and underlying questions (see Table 5.1). Here, as perhaps in other disciplines, individual representatives who represent certain systematisations can only be named to a limited extent. Individual publications are therefore used for classification rather than the authors themselves. It is striking that in all systematisation approaches time and historical changes play a role (either explicitly as a distinguishing feature or implicitly in the representations of the planning theories as a diagram with a time axis). The distinction between procedural and substantial theories is at least an implicit basis in some structuring approaches and is expressed in the focus on procedural approaches.

Historical perspective A systematisation approach that is fundamentally based on the chronological classification of planning theories is the close linking of historical social development with planning theories or so-called paradigms of planning. A model that is widely known in German-speaking countries is the combination of the phase model according to Albers (1993) and the layer model according to Selle (1995). A distinction is essentially made between four phases and four layers,

Tab. 5.1 Systematisation approaches of planning theories (own illustration)

Perspective	Goal		Literature
Historical perspective	Changes in the understanding of planning over time	• Temporal course and change as a transition between planning theories	Selle (1995), Albers (1993), Fainstein and Campbell (2012)
Perspective of separation of content and process	Distinguishing between substantial and procedural planning theories	• Classic separation • in theories *in* and *of* planning	Yiftachel (1989), Faludi (1976)
Perspective of planning generations	Planning as solving (complex) problems over time	• Extensive changes in the scientific-theoretical foundations	Schönwandt (2008)
Perspective of planning theories as an expression of collective action	Planning as an act of government in a much broader democracy	• Very broad and comprehensive systematisation	Friedmann (1987)
Perspective of the fields of application in practice and research	Differentiation of planning theories based on tasks for practice or research	• Tasks that are important in different fields of application	Beauregard (2020), Rydin (2021)

the first-mentioned phases are divided into adaptation planning, containment planning, development planning and perspective planning over time, with the layers underneath to outline that the respective planning contents and understandings are not replaced, but rather new layers are added and planning in this sense becomes more complex in the direct sense of the word (see Fig. 5.1). Similar to the phases, Selle (1995) divides the shifts into hazard prevention, setting the framework for private construction work, planning coordinated developments and acting cooperatively. Wiechmann also introduces planning theory (Wiechmann 2019a, 2019b) on the basis of the historical development of planning in planning theories. Starting with lines of development since the 1950s, he traces the change in planning theories in the form of four "big shifts", "from administrative-technical planning to a task for society as a whole, from the scientific search for optimal solutions to collective learning processes, from the interventionist claim to control to communicative action and from planning vicarious agents to politically active planning actors" (Wiechmann 2019a, 2019b, p. 8, own translation). These shifts can also be classified in terms of time, even if they have a different scientific meaning than phases or shifts. The historical classification implicitly includes an ontological turn, namely that of realism (in rationalist planning) towards constructivism (in cooperative planning), which is often treated in the form of paradigms in planning sciences. Epistemologically and theoretically, the phases or layers build on each other, only an *accumulation* of knowledge is differentiated, without separation of knowledge dimensions. This form of systematisation is heavily concerned with the genesis of these theories, less with their validity. Roughly speaking, the question of genesis explains how a certain position comes about and the question of validity explains whether this position is possibly true.

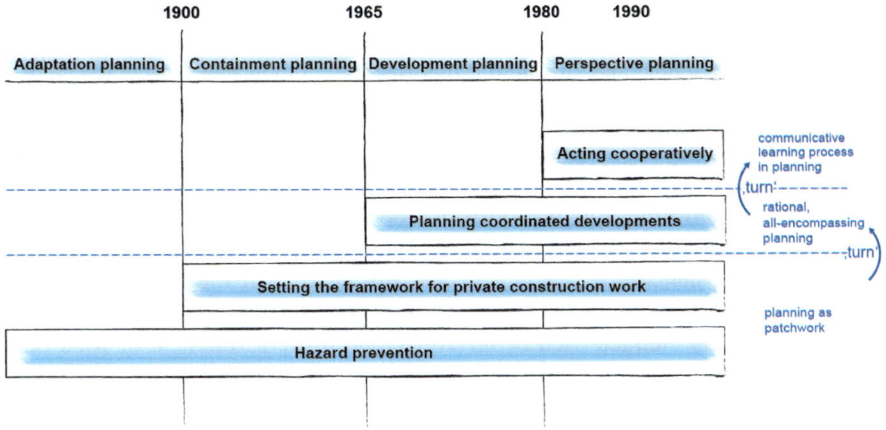

Fig. 5.1 Historical perspective on planning theories. (According to Selle 1995, modified according to Wiechmann 2019a, 2019b, own illustration)

Perspective of the separation of content and process Other systematisation approaches put the distinction between substantive and procedural theories in the foreground, i.e. the distinction between content and processes of spatial planning (and possibly the context). Faludi (1976) was the first to publish this differentiation of theory in planning and theory of planning. Theories in planning relate to substantive theories within the sub-areas of planning, such as land use, transport, urban planning, regional development, environmental planning, etc. (Friedmann 2003, p. 7). These theories deal with what is at stake, how infrastructures should be distributed in space, what constitutes a sustainable city or how streetsshould ideally be structured (Allmendinger 2009; Friedmann 2003). Planning theories deal with how the process of planning is or should be implemented and serve to structure and analyse the planning process itself. These procedural theories place the procedural course in the foreground of planning perspectives and include formal and informal consensus-building, decision-making and implementation processes as well as the assertion of interests through power, conflict and consensus (Wiechmann 2019a, 2019b, p. 4). It is interesting here how the future orientation, which also occurs in the knowledge dimensions (status quo and aspired state), can be found here. In the planning-theoretical discussion, the process dimension is of outstanding importance; it has been the focus of theoretical reflection for decades. Procedural planning theories are often understood under the narrower definition of the term planning theory. Even if the authors named here not only differentiate between content-related and procedural planning theories, but also define other categories, they can be significantly distinguished from other systematisations under this extremely characteristic subdivision. Yiftachel (1989) for example, distinguishes theories into the "analytical debate" (content), the "urban form debate" (normative theories) and the "procedural debate" (process), while Wiechmann (2019a, 2019b) deepens the division into context, content and process. The result of this systematisation, which has been criticised for 50 years now, is that planning theories and their discussion are predominantly expressed in procedural approaches, and substantial theories have fallen behind. Understanding planning on a theoretical level primarily as a procedural enterprise or considering and setting up theories from this perspective prevents a holistic perspective that goes beyond the (important) procedural elements of planning theories.

Perspective of planning in generations In order to categorise planning approaches, the first, second and third generation of planning has been discussed since the 1970s. The first generation was taught and practiced until the early 1970s. This first generation of planning is largely based on the rationalist planning model and is accordingly characterised by an overly optimistic assessment of what planning can achieve: "there are no problems, only solutions" (Catton 1980, p. xiii). According to this view, planners are capable of making rational and objectively comprehensible decisions that precede their planning actions (Schönwandt 2008, p. 20). Rittel (1972), who introduced the classification according to generations, contrasts the first generation with a second generation. Above all, he sees a distinction in the consideration of problems. In the first generation, the focus was on solvable problems and in the second generation on wicked problems. Problems that can be solved are, for example, solving mathematical equations (Schönwandt 2008, p. 20), the solution path and the solution to be achieved are clearly defined. Wicked problems (Rittel 1972) are much more complex, have no defined solution path, and are essentially unique. There is no way to learn through trial and error and no way to finally verify the solution. This approach and perspective made waves. Above all, the introduction of scientific-theoretical basic assumptions in the field of planning knowledge should be mentioned—such as the realisation that all knowledge is surrounded by uncertainties and that correspondingly objective problem descriptions and solutions are not possible. Schönwandt (2008) introduces a third generation, which is based on a simplified presentation of Luhmann's systems theory. Here, he primarily uses the System-Environment Paradigm, which assumes a system core that is in turn embedded in an environment. Accordingly, systems theory is primarily a theory of relationships, of networks between the system core and the respective environment. Only those aspects that are connected to the core and not to the general system are understood as the environment (Schönwandt 2008, p. 27). In addition, Schönwandt attempts to get away from the concept of *planning* and to present planners as the individuals with their specific biological and psychological characteristics, being interwoven in cooperation and institutions and who live and work in a specific social and cultural environment (Schönwandt 2008, p. 27). The System-Environment Paradigm mentioned is already fundamental for the functional circle of Jacob von Uexküll's biologist. Schönwandt (2008) transfers this to planning: He assumes two worlds—the "life world" and the "planning world". The latter is embedded in the first and both are interrelated. The cycle through these two "worlds" is crucial for the model. While the "planning world" covers the area in which plans are created and planners work in various "institutions" (for Schönwandt also organizational forms), the "everyday world" includes the entire environment. Accordingly, the actors in the planning world and their value judgments when selecting approaches have a strong influence on the results of planning. The conceptual pairing agenda and arena is central here. The agenda includes the political discussion and dispute points, which in turn can be the starting point for planning projects and actions. The arena, on the other hand, represents the total number of actors who have different interests and options for action and who influence the discussions (such as citizens, associations, authorities, companies, etc.). The arena-agenda constellation thus reflects all the "economic, political, social and ecological, circumstances of a planning task" (Schönwandt 2008, p. 37). The enhancement of planning-theoretical approaches in generations reflects the leaps from a mono-rationalist perspective (first generation) to a change in the view of the object of knowledge to idealism (second generation) to a social-constructivist perspective in the third generation.

Perspective of planning theories as an expression of collective action One of the most far-reaching attempts of a systematisation in which planning theories are only part of the field is that of Friedmann (1987, *Planning in the Public Domain*). He positioned planning at the intersection of reason and democracy, in which planning attempts to develop future possibilities from existing conditions. His understanding of planning is broad, based on collective action, which includes planning in the narrower sense and is in some way related to state, public action. While other systematisations regard planning as a process, Friedmann's systematisation focuses on *planning as an idea*. He presents a typology of four different approaches to planning theories, Policy Analysis, Social Learning, Social Mobilization and Social Reform. All four types are described in detail, based on philosophical schools of thought and thinkers. Political analysis as scientific rationality, which advises political decisions, and social reform, which identifies social problems and proposes solutions for them, are interesting for spatial planning in the narrower sense. In this model, the historical-theoretical classification of the respective planning-theoretical positions is reproduced in detail. Likewise, connections between the different approaches become clear in their theoretical development. At the same time, this representation is not sufficient if the current state of planning theory is to be analysed, since it primarily refers to the historical development and thus primarily the retrospective is highlighted.

Perspective of the fields of application in practice and research Recently, some approaches to the systematisation of planning theories were published, which in a certain respect sort them according to fields of application of the planning theories: On the one hand planning theories specifying the tasks for people in practice, on the other hand handbooks making the tasks for planning research explicit. In 2020, Beauregard published a new systematisation of planning theories, a systematisation based on the planning tasks Knowing, Engaging, Prescribing, Executing. He himself writes that he is limited to procedural planning theories and that substantive theories hardly play a role and that his systematisation represents a strongly Anglo-American perspective. Authors and theories from other language areas and continents are not considered. What is interesting in the description of the individual planning tasks is which aspects come into play in them. With this systematisation, he draws heavily on theories of practice and theories of social learning (Beauregard 2020, p. 14), i.e. theories that relate to social relationships rather than overarching structures. He emphasises the focus on planning tasks, which are not to be classified in a chronological order, but as a heuristic framework that makes it possible to compare and contrast the different planning theories (Beauregard 2020, p. 15). His focus on planning tasks gives rise to the thesis that Beauregard's approach is primarily addressed to people in planning practice. Another perspective that represents more of a systematisation of planning research approaches is the book Theory in Planning Research by Yvonne Rydin (2021). It focuses on the relationship between the way we think about the world (theory) and the way we study it (research). In doing so, she systematises the field of planning theories according to their content for research into planning practices, from the perspective of researchers who are looking for a suitable approach to answer their research questions. It provides a valuable guide to dealing with planning theory for students and researchers to choose a theoretical perspective. However, the narrowing of the theories she chooses is a result of her personal experiences and interests:

Government (Governmental Models), Rational-choice Perspectives, New Institutionalism, Governance Theories, Urban Politics, Political Economy, Discourse and Knowledge, and Relational approaches. What is valuable about this presentation, which represents a valuable collection of theories, is the disclosure of the basic premises and theses of individual theories and the classification of which research questions can be answered with them. This concrete representation of the structure of individual theories and their knowledge perspective is not to be confused with a systematisation, but it represents a valuable, intensive examination of what makes planning theories.

Implications for planning theories—first conclusions As already mentioned, all these systematisations have a purpose on which they build. This makes clear that none of the systematisations is wrong or correct, but that they are more or less useful for orientation and classification of planning-theoretical approaches. The historical systematisation, for example, is extremely useful for clarifying that planning theories depend heavily on the societal understanding of a state and are therefore extremely adaptable. The focus on the genesis of planning theories can also be easily combined with the history of spatial planning, an approach often used in academic teaching. The historical systematisations seem to be less useful if e.g. rationalist planning is assigned to a period of time and conveys that there is no longer any rationalist planning today, since it was superseded in the 1970s. The separation of process and content, on the other hand, was groundbreaking for the further development of planning sciences and still shapes its orientation to a large extent today. On the other hand, the substantive theories and their theoretical foundation have tended to be neglected, since the literature deals almost exclusively with procedural theories. The division has led to a dichotomisation of these two areas, although many authors emphasise the interdependence of the dimensions. The future orientation of planning makes content without process thinking obsolete, but even an isolation of the processes without content remains empty. However, these are not the only difficulties associated with the relatively strict division into procedural and substantive theories. In a certain respect, this subdivision has become a paradigm within the planning sciences, which significantly shapes the perspective on theories and hardly allows other perspectives. In the discussion about planning theories, procedural theories are dealt with almost exclusively, while substantive ones are sparsely considered. This also results in the difficulty of defining substantial theories at all, since a scientific language is missing insofar as procedural theories are primarily dealt with and the terms and discussions are therefore geared towards them. The consideration of the process dimensions of planning has become independent to a certain extent, has almost become an objective in itself. Such a limitation of the perspective is basically unproblematic, however, the separation of content and process, as this dichotomy suggests, is not a tenable theoretical position that does justice to the diverse field of planning knowledge. Planning processes are primarily a means of conveying certain content-related objectives and are not an objective in themselves, even if complex bureaucratic processes and structures can certainly give a different impression. The closely interlocked consideration of these two elements, the respective framework and the content that fills it, is correspondingly necessary. Especially since both are in constant exchange and influence each other. Depending on which content-related perspective, which theoretical conviction is followed, this can have serious consequences for the framework, i.e. the instruments, the means and the ways of implementation. Framework and content are thus significantly less rigid and are in much more exchange than the strict division into procedural and substantial allows for. Depending on the content, the frame can be deformed or is deformed by it and adapted to the respective content-related positions. For example, planning based on the concept of Just City pursues completely different goals and content than planning based on the concept of incrementalism and, as a result, formulates completely different ideas about what the framework for achieving the respective goals and content looks like. The means are adapted to the objective and not the other way around.

Implications for planning theories—further approaches The introduction and identification of generations of planning can be seen as a further development of the systematisation, although it insists on a historical classification on the one hand, on the other hand the three generations also refer to fundamental epistemological differences: between realism and social constructivism, between the subject of knowledge and the object of knowledge. However, the importance of knowledge and the ethical component remain at least hidden. Friedmann's systematisation of planning theories opened perspectives on what the idea of planning could be. His comprehensive analysis of planning theories and their typification as collective action were decisive for the epistemological classification in philosophical schools of thought. This also corresponds to the level of reflection that depicts the types of planning theory. On the other hand, there are more up-to-date representations of planning theories, as aspects and tasks dedicated to practice or research. Here, different partial aspects of an epistemological consideration are deepened and presented, always with the aim of better, easier handling in the fields of application in practice and research. This does not represent a systematisation directly, but it does mean a scientific-theoretical enhancement. Current approaches to planning theories and even systematisation have tended towards post-positivism, i.e. diversity in a complex, indefinable juxtaposition. The field of planning theories no longer provides orientation, it is more a matter of "picking and choosing" individual theories than of classifying what planning and planning theories can be. Systematisations can give a clear picture of seemingly incomprehensible debates on planning theory and contribute to a better understanding of their meanings. Or, as Allmendinger (2009) states: "Planning theory now has a diverse and fragmented landscape. The need for a typology to help organize and explain these positions in relation to different schools of theory, other disciplines and planning practice is as necessary as ever."

References

Albers G (1993) Über Den Wandel Im Planungsverständnis. Raumplanung 61:97–103

Alexander ER (2010) Introduction: does planning theory affect practice, and if so, how? Plan Theory 9(2):99–107. https://doi.org/10.1177/1473095209357862

Alexander ER (1997) A mile or a millimetre? Measuring the 'planning theory—practice gap. Environ Plann B Plann Des 24–31. 3/b240003

Alexander ER (2016) There is no planning—only planning practices: notes for spatial planning theories. Plann Theor 15(1):91–103. https://doi.org/10.1177/1473095215594617

Allmendinger P, Tewdwr JM (1997) Mind the gap: planning theory—practice and the translation of knowledge into action-a comment on Alexander (1997). Environ Plann B Plann Des 24(6):806–809. https://doi.org/10.1068/b240802

Allmendinger P (2002) Towards a Post-positivist typology of planning theory. Plan Theory 1(1):77–99. https://doi.org/10.1177/147309520200100105

Allmendinger P (2009) Planning Theory. Palgrave Macmillan, Basingstoke

Alterman R (2017) From a minor to a major profession: can planning and planning theory meet the challenges of globalisation? Transactions of AESOP 1(1):1–17. https://doi.org/10.24306/TrAESOP.2017.01.001

Barker G, Kitcher P (2014) Philosophy of science—a new introduction. Oxford University Press, New York

Beauregard RA (2020) Advanced introduction to planning theory. Elgar, Cheltenham

Birch EL (2001) Practitioners and the art of planning. J Plan Educ Res 20(4):407–422. https://doi.org/10.1177/0739456X0102000403

Brehmer M (2003) Wissenschaftstheorie. In: Wulff DR (ed) Handwörterbuch Philosophie. Vandenhoeck & Ruprecht (UTB Philosophy, 8208), Göttingen, pp 685–688

Bunge M (2003) Philosophical dictionary. Enl. ed.Prometheus Books, Amherst

Bunge M, Broermann C, Murguia A (1983) Epistemologie Aktuelle Fragen der Wissenschaftstheorie. Bibliographisches Institut, Mannheim

Catton WR (1980) Overshoot. The ecological basis of revolutionary change. University of Illinois Press, Urbana

Davoudi S (2015) Is planning an academic discipline? RIURBA (Numéro 1). http://www.riurba.review/Revue/is-planning-an-academic-discipline/. Accessed 22 July 2021

Davoudi S, Pendlebury J (2010) Centenary paper: the evolution of planning as an academic discipline. Town Plann Rev 81(6):613–646. https://doi.org/10.3828/tpr.2010.24

de Neufville JI (1983) Planning theory and practice: bridging the gap. J Plan Educ Res 3(1):35–45. https://doi.org/10.1177/0739456X8300300105

Fainstein SS, Campbell S (2012) Readings in planning theory. Blackwell, Oxford

Faludi A (1976) Planning theory. Repr. Pergamon Press (The urban and regional planning series, 7), Oxford

Friedmann J (1987) Planning in the Public Domain. Princeton University Press, New Jersey; Friedmann J (2003) Why do planning theory? Plann Theor 2(1):7–10. https://doi.org/10.1177/1473095203002001002

Friedmann J (2003) Why do planning theory? Plann Theor 2(1):7–10. https://doi.org/10.1177/1473095203002001002

Friedmann J (2008) The uses of planning theory. J Plan Educ Res 28(2):247–257. https://doi.org/10.1177/0739456X08325220

Gualini E (2010) Planungstheorie. In: Henckel D, Kuczkowski K, Lau P, Pahl-Weber E, Stellmacher F (Hrsg) Planen – Bauen -Umwelt. Ein Handbuch. VS Verlag/Springer Fachmedien, Wiesbaden, S 365–369

Hemmens GC (1980) New directions in planning theory introduction. J Am Plann Assoc 46(3):259–260. https://doi.org/10.1080/01944368008977041

Hopkins LD (2001) Planning as science: engaging disagreement. J Plan Educ Res 20(4):399–406. https://doi.org/10.1177/0739456X0102000402

Kuhn TS (1962) The structure of scientific revolutions, 4th edn. The University of Chicago Press, Chicago

Lord A (2014) Towards a non-theoretical understanding of planning. Plan Theory 13(1):26–43. https://doi.org/10.1177/1473095213477642

Pinson D (2004) Urban planning: an, undisciplined‘ discipline? Futures 36(4):503–513. https://doi.org/10.1016/j.futures.2003.10.008

Resnik DB (2000) A pragmatic approach to the demarcation problem. Stud Hist Phil Sci 31(2):249–267. https://doi.org/10.1016/S0039-3681(00)00004-2

Rittel H (1972) On the planning crisis: systems analysis of the, first and second generations'. Bedriftsøkonomen 8:390–396

Rydin Y (2007) Re-examining the role of knowledge within planning theory. Plann Theor 6(1):52–68. https://doi.org/10.1177/1473095207075161

Rydin Y (2021) Theory in planning research, 1st ed. Palgrave Macmillan (Planning, environment, cities), Singapore

Schönwandt W (2008) Planning in crisis. Theoretical orientations for architecture and planning. Routledge, New York

Scott AJ, Roweis ST (1977) Urban planning in theory and practice: a reappraisal. Environ Plan A 9(10):1097–1119. https://doi.org/10.1068/a091097

Selle K (1995) Phasen oder Stufen? Fortgesetzte Anmerkungen zum Wandel des Planungsverständnisses. RaumPlanung 71:237–242

Strohschneider P (2017) Über Wissenschaft in Zeiten des Populismus. Forschung—Mitteilungen der DFG 42(3):I–VIII. https://doi.org/10.1002/fors.201770312

Wiechmann T (Hrsg) (2019a) ARL Reader Planungstheorie, Bd 1. Springer, Berlin

Wiechmann T (Hrsg) (2019b) ARL Reader Planungstheorie, Bd 2. Springer, Berlin

Wiesen H (2003) Theorie. In: Wulff DR (Hrsg) Handwörterbuch Philosophie. Vandenhoeck & Ruprecht (UTB Philosophie, 8208), Göttingen, pp S645–646

Yiftachel O (1989) Towards a new typology of urban planning theories. Environ Plann B Plann Des 16(1):23–39. https://doi.org/10.1068/b160023

Topology of Planning Theories

6

Abstract

The Topology of Planning Theories as a newly introduced systematisation builds on the characteristic features of planning knowledge: the location in the field of knowledge and the degree of action-orientation of planning theories. The topology outlines the diversity and importance of planning knowledge for the profession and discipline. The field of planning knowledge contains epistemological foundations, different theories each find their specific place. Basic classification is based on the question of how reality(ies) are grasped and what knowledge is contained in the theory.

Keywords

Topology of planning theories • Field of planning knowledge • Action orientation

Need for a different systematisation Spatial planning and its understandings are diverse, as are the theoretical perspectives that can be taken on it. Systematisations of theoretical approaches generally help to identify significant differences, but also similarities between individual theories and to relate them to each other. Science differs from knowledge in that it is a system of true statements (Poser 2012, p. 20). Since the order of the sentences or statements is systematic and well thought out, not every formulated sentence is already science. Comparable to a matrix or a fabric of sciences, the purpose of a systematisation is not the demarcation and exclusion of individual theoretical contributions, but rather the basis of each science for orientation and direction in the field of science. The extent to which spatial planning is actually a science or not is a matter of debate in some publications (Behrend and Levin-Keitel 2020). The basis of this discussion is always the question of how planning knowledge can be understood, how it can

M. Levin-Keitel and L. Behrend, *The Topology of Planning Theories*, The Urban Book Series, https://doi.org/10.1007/978-3-031-37857-7_6

be differentiated from everyday knowledge and how it can be justified–according to Hoyningen-Huene (2013) a question of the systematicity of science as a differentiation from largely unstructured everyday knowledge.

The dominance of the divide into procedural and substantive planning theories The very prominent presentation in planning sciences of *theories in planning* and *theories of planning* (Faludi 1973), i.e. into substantive and procedural planning theories, represents such a systematisation. The distinguishing feature is the research object, i.e. the question of the content of spatial planning or the process of how future developments are to be achieved. In planning-theoretical discussions, the process dimension is of outstanding importance; it has been the focus of theoretical reflection for decades. Although it is always emphasised that one dimension should not be viewed separately from the other, in the strong focus on procedural theories in research and science, substantial theories hardly appear. Accordingly, this leads to a certain restriction on what constitutes planning, if the procedural level is presented relatively dominantly in comparison to dealing with the content dimension of planning. Similar to the historical presentation of the theoretical development of planning theories, the pure consideration of the procedural dimensions cannot make any statements about the validity of the respective theory–it remains primarily descriptive in nature. The Topology of Planning Theories takes on this basic idea by introducing a systematisation that reduces planning theories to their basic components. With the attempt to break down the different epistemological aspects of planning theories and thus to describe theoretical fields in planning sciences, the modes of action and validity claims of the theories are outlined and existing dichotomous representations are at least weakened.

6.1 Basics of a Topology–From Genesis to Validity

Change of perspective from genesis to validity The Topology of Planning Theories is not based on a historical classification of planning theories. Instead of dealing with the genesis of the theories (an important, and often applied systematisation), the Topology of Planning Theories deals with their validity. Accordingly, it focuses on its justifications and their areas of validity, which are represented in planning theories in form of certain viewpoints and perspectives. Instead of dealing with their emergence and historical development, the current planning-theoretical *perspective* is analysed and the knowledge inherent in the theories is characterised. An important component is the subdivision of the theories into their specific perspectives on reality(s), present and future as well as normative orientations, which becomes visible through the topology. We aim to answer rather questions like: Which understanding of planning does the respective theory require, which assumptions are made about valid forms of knowledge and which normative framework is set? Relevant questions when deciding on a meaningful application and consideration of planning theories. This more in-depth examination of planning theories and their epistemological foundations has hardly taken place to date.

The result is a barely justified differentiation of planning theories in research and practice. Accordingly, the Topology of Planning Theories aims at a systematisation of this inherent knowledge and its location in the field of planning theories.

Structure of the topology The Topology of Planning Theories is a systematisation of planning-theoretical approaches that is neither marked by right nor wrong, but meets the scientific requirements of a disciplinary foundation that understands planning knowledge as a scientific field and places planning theories in relation to one another. The term topology emphasises the systemic nature of this scientific field and how certain elements relate to each other; not to be confused with a typology with the goal of a distinction and differentiation between individual elements. Deviating from a linear historical genesis or a cause-effect logic, interactions and connections between planning theories are discussed here, but also their differentiation from each other. Rather, the Topology of Planning Theories is a logic of locating theories, their objects of knowledge, scope, limitations and characteristics. This classification is not linear or even causal, the interactions are more like double images or patterns that fit together. The special feature of the planning sciences lies both in their current and future-oriented planning knowledge, as well as in their strong action-orientation on planning practice and a directly corresponding field of action, spatial planning. The Topology of Planning Theories picks up on these fundamental characteristics of the field of planning knowledge and uses them to describe planning theories and their theoretical fields in more detail:

- **Positioning in the field of planning knowledge** Planning theories can be differentiated in terms of their reference to reality and the types of knowledge they contain, from rationalist perspectives to a social-constructivist worldview, with descriptive-analytical knowledge about the status quo or future-oriented, normative knowledge. These descriptions of knowledge serve as a fundamental field of planning theories, in which different theories can be related to one another (Sect. 6.2).
- **The degree of action orientation of planning theories** The action-orientation of planning theories has traditionally had a very high status in planning sciences, with different planning theories playing different roles in their application: from actually action-guiding theories to social-theoretical approaches that place spatial planning in a societal context (Sect. 6.3).

6.2 The Field of Planning Knowledge–An Epistemological Systematisation

Knowledge description of planning theories The field of planning knowledge can be defined significantly on the basic assumptions of the respective planning theory, i.e. what constitutes the inherent understanding of planning (cf. Chap. 2), which knowledge base is considered, i.e. how does a planning theory recognise which kind of knowledge (cf. Chap. 3) and based on the degree of normativity, i.e. what planning should do according to a certain theory (cf. Chap. 4). Each of the three scientific-theoretical dimensions form different planning theories and enables a differentiated localisation in relation to other theories and scientific approaches. The three dimensions are closely linked in the theories and often result from each other. So, it is obvious that a consideration of future-relevant knowledge specifies a strongly normative orientation, because the desired future is already setting up values to achieve: Whether it's the green, healthy city or a collaborative process in which the future is shaped. This leads to an integrated consideration of the epistemological and ethical dimensions, a peculiarity resulting from the types of knowledge in planning sciences. Locating theories in the field of planning knowledge represents then the specific perspective of the respective theory, its inherent knowledge, its opportunities, but also its limits and threats. Locating planning theories in the field of planning knowledge is then, on the one hand, largely determined by the question how many realities are recognised and how they are dealt with; on the other hand, planning theories can be characterised on the basis of their inherent stock of knowledge and the respective ethical-normative orientation (see Fig. 6.1).

Planning Theories as an Expression of a (Variety of) Reality (ies)

One or more realities Ontology deals with a number of basic questions, e.g. questions of the existence, of the being and the becoming, or of what we call reality. With regard to planning theories, basic ontological positions become interesting when it

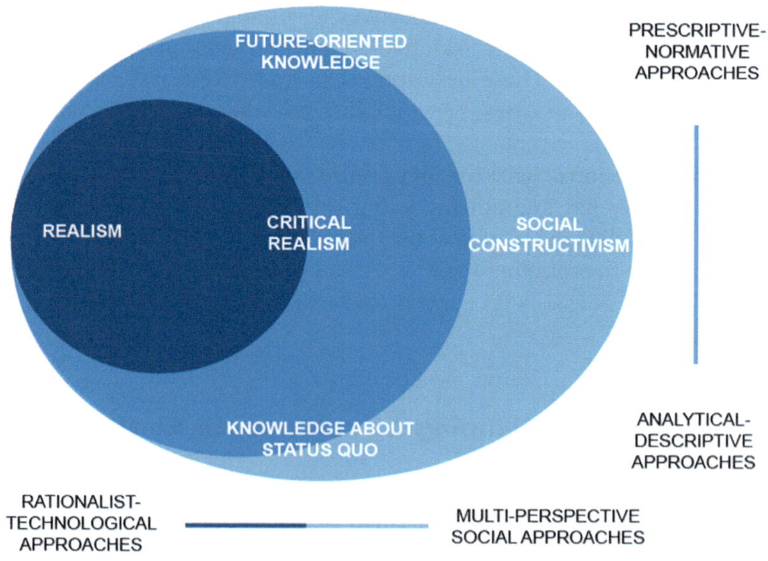

Fig. 6.1 The field of planning knowledge—an ontological description of knowledge. (own illustration)

comes to the perspective of recognised realities, i.e. planning practices. Questions of this dimension relate to how many realities are (not) recognised in the respective understanding of planning, how spatial planning deals with other realities and who, how and what is planned (cf. Chap. 2). Decisive is then, the acceptance of one or more *realities* of planning and the recognition of other (planning) realities and practices. Here, derived from ontological and epistemological understandings of science, two different theoretical fields can be identified, rationalist planning theories (associated with realism) and socially-constructivist planning theories (based on social constructivism). The shape of planning theories thus reflects the relationship between built space or technology–i.e. one reality of planning–and individual or social behaviour or societal consequences. Classically, the theories of rationalist planning and communicative planning, often described as paradigms in planning sciences, describe two extremes in the continuum between one and several acknowledged realities.

Rationalist approach and objectivity If the theory assumes one reality that is measurable and adequately describable, the planning theory positions itself close to realism and emphasises a rationalist understanding of planning. The theoretical framework is relatively clearly outlined within a rationalist positioning, the planner collects all available and required information and uses (apparently objective) logic to argue towards a decision. It is the "ratio", i.e. an intellectual achievement, on which decisions are made and which legitimises them. The foundations of decision-taking must therefore be as objective and clear as possible: calculations and objective measurements, such as density, accessibility, socio-demographic surveys or area percentages, which can usually be expressed in numbers, are particularly suitable. The ontological form reveals itself in the question of which reality is represented and in particular, how the desired state can be (objectively) described and made measurable. In terms of content, a rationalist perspective is expressed in theories such as the 15-min city or spatial typologies based on population densities, for instance. While the former are based on knowledge of accessibility and can thus be measured in distances and the time required, the latter is due to the subdivision of town and country according to population density.

Multi-perspective approaches and social-constructivism From a social constructivist point of view, however, planning theories introduce further perspectives on reality going beyond one's own perception of the one reality and requiring to deal with multiple perspectives on reality. Planning theories that can be positioned here–and this includes the majority of modern and postmodern planning theories–require dealing with different constructions of reality and understand planning strongly as a social process of negotiating different perspectives of reality. This social-constructivist perspectives are represented by planning theories such as communicative planning, which include a large number of actors and thus perceptions of realities, which uncover these realities in joint (communication and collaboration) processes and must be taken into account in planning. The degree of recognition of these multiple realities varies, from acknowledgment to the development of a common conception of reality as a basis for decision-making and planning. In terms of content, theories such as spatial justice should be mentioned here, which relate to social justice issues and different constructions of reality. Methodologically, these must be captured and socially discussed, in participatory processes by joint learning and decision-making.

Forms of Knowledge of Planning Theories

Knowledge of the present and the future The field of planning knowledge can also be described by classifying the forms of knowledge inherent in the theories. Not all planning theories contain the forms of knowledge mentioned, which is why it is very helpful to classify the type of knowledge, especially for the scope and the object of investigation and thus the understanding of planning. Theories, seen as organisation of knowledge and the relevant basis of a discipline, contain different types of knowledge. In the planning sciences, these are four fundamental forms of knowledge: Knowledge from information and facts (descriptive), knowledge from the analysis of these facts (analytical), knowledge about the future (prescriptive) and knowledge about how planning ought to be (normative) (cf. Chap. 5). It should be pointed out that descriptive-analytical planning theories also contain a more or less strong focus on an objective or a statement that follows from the description or the analysis. Nevertheless, planning theories can be assigned to the different forms of knowledge in terms of the focal points and goals they address. As has already been made clear, ethical aspects play a decisive role in the epistemological dimension: Especially future-oriented knowledge, i.e. knowledge to achieve a desired state in the future, is highly normative. If the theory formulates a desired future, i.e. if it formulates an *ought*, it must make statements about which desired future is to be achieved (normative) and act in a predictive (prescriptive) manner. At this point we refer to the naturalistic fallacy and the statement that there is no direct way from an *is* to an *ought* (from an analysed state about *what is* to a future imagination *what ought to be*), from a status quo to determining how something should be–this succeeds only through a normative definition of whatever kind, i.e., an evaluation of the status quo in favour of a desired future.

Descriptive-analytical forms of knowledge Descriptive-analytical forms of knowledge contain knowledge that deals with the collection of information and facts (empirical knowledge or descriptive knowledge) and knowledge in the sense of an analytical consideration of the facts at hand. They refer to knowledge about the status quo, can either describe it (descriptive) or look for (causal) relationships (analytical). Examples are the actor network theory, the concept of ecosystem services or the analytical understanding of governance, as in actor-centred institutionalism. Descriptive-analytical planning theories from other disciplines are often introduced into the planning sciences, mostly from geography, economics, political and social sciences or other disciplines that are particularly strong in these forms of knowledge. Even if the strong future orientation of the planning sciences is reflected more in the prescriptive-normative types of knowledge, this descriptive-analytical knowledge serves as a decisive basis for planning the future.

Prescriptive-normative forms of knowledge Prescriptive-normative forms of knowledge of a state to be achieved in the future characterise planning theories to a particular degree: This future orientation is fundamental for planning theories, it distinguishes planning sciences from other disciplines such as its sister discipline, geography. There are many examples, e.g., the 15-min city, the garden city or communicative planning, Marxist planning, etc. Since knowledge about future processes or states always also contains knowledge about the status quo as the basis for future orientation, the epistemological classification can be presented in levels rather than in a continuum. The 15-min city builds on the current accessibility and car dependency, the garden city served as an answer to urban deficiencies and social objectives, communicative planning stands in contrast to rationalist planning and its weaknesses and limits, Marxist planning defines planning in a different political system, characterised by the critique of capitalism. In determining the forms of knowledge, it is not a question of either/or, but rather of setting the focus of a theory.

6.3 The Degree of Action-Orientation in Planning Theories–Macro, Meso and Micro Level

Inclusion of the characteristic action-orientation A further aspect that makes it necessary to introduce the topology of planning as a new systematisation is its ability to depict the different levels of action-orientation in planning theories. The communicative planning theory represents a way in which planning and planning processes should be understood and exercised as a social task, while e.g. incrementalism has a much stronger action reference and has a direct action-guiding character. The special characteristic of planning theories, compared to other disciplines, is represented in a strong action-orientation in general. This action-orientation coins out on different theoretical levels, serving as a main differentiation in the Topology of Planning Theories: planning as an expression of a societal decision-making (social theories), planning as an expression of a specific activity and task in a society (holistic planning theories) and planning action as an expression of specific activities to influence or control the future (action theories) (see Fig. 6.2).

Macro, meso and micro level The differentiation of action-orientation in planning theories into the social macro level, a (neo-)institutional meso level and an action-oriented micro level mirrors both the narrow and broad understandings of

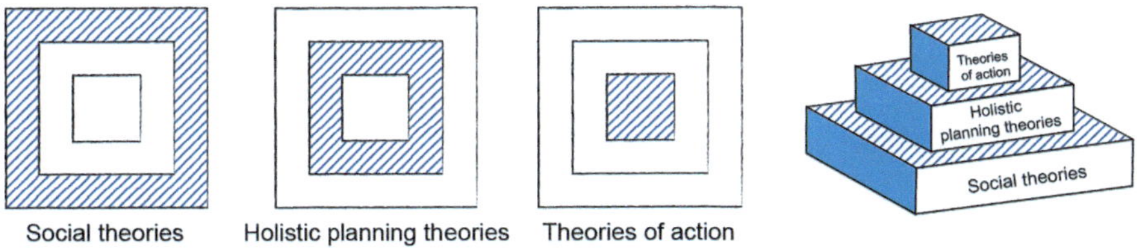

Fig. 6.2 The degree of action-orientation in planning theories–macro, meso and micro level (own illustration)

planning (i.e., who is seen as a planner), as well as different disciplinary backgrounds (i.e., the theory has roots in social science, political science or economics, for instance). It seems that the disciplinary roots of a respective planning theory decisively define the respective level of action. Theories from social sciences like the discourse theory by Habermas relate to planning as a social task, political theories reflect planning issues from a political perspective, theories from management sciences are much more concerned with action-oriented reflections and address planning practice directly (see Table 6.1). The differentiation into narrow or broad planning understandings determines the action-orientation, since it is crucial who should carry out these actions. While we mainly assume a narrow understanding of planning in procedural theories (planning as state action), this must first be further defined in substantive theories. For instance, the spatial justice theory initially says nothing about the role that governmental planning should play in this: Is the planning department an advocate of spatially just development? Or does the theory aim more at the disadvantaged and their rights to self-determination? Then, spatial planning could even be seen as one of the causes of a spatially unjust developments.

Social Theories

Planning as part of society Planning theories that stem from or can be derived from a changed understanding of the state, are referred to in the Topology of Planning Theories as social theories. In planning sciences, social theories serve to show how planning is embedded in an overall social system. Strongly influenced by normative knowledge elements, the question here is how we would like to live as a society. Social

Table 6.1 Action-orientation and inherent differences at macro, meso and micro level. (own illustration)

Social theories	Holistic planning theories	Theories of action
		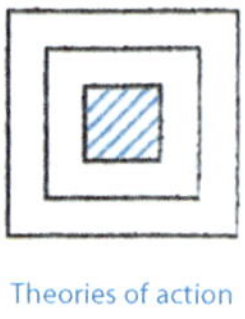
Societal macro level	(Neo-)institutional meso level	Action-oriented macro level
• Transferred from social sciences and humanities • What should our society look like?	• Rather from cultural sciences, management sciences, political sciences • What does planning look like? Who is planning, what does the context look like, etc.?	• Rather from management sciences, economics, etc • What can planning actually do and how?
• Broad understanding of planning, understanding of decision-making in society	• Broad understanding of planning, role of planning in society	• Narrow understanding of planning and what planning is supposed to do

theories relate to society as a whole and try to conceptualise central features of society. For instance, the ideas of Marxism determine a fundamentally different role and a different understanding of spatial planning than Habermas' discourse theory. Social theories are always a source of inspiration, also for the planning sciences and planning itself, in its task as a public sector to anticipate in shaping social challenges–the city of tomorrow, the future of a region. In most cases, social-theoretical approaches are only partially transferred to planning sciences, according to the principle of "best fit" or "pick and choose". Planning theories are more embedded in these social theories or, represented as the basement of holistic planning theories in the pyramid, build on them as basic understanding (Fig. 6.2). The challenge here is the often implicitly outlined utopian character of social theories, such as what a Marxist state should look like and what role planning then has or how consensus-oriented planning should be implemented in politics (discourse ethics).

Social theories are relevant to planning sciences for several reasons:

(1) They describe the society or a utopian idea thereof, in which planning processes are organised and institutionalised. Depending on the reference to a social theory, the goals, content and methods of the public sector also change. Against the background of discourse ethics (Habermas), decision-making processes differ significantly from rationalist planning processes–especially with regard to participation approaches or the social legitimacy of decisions.
(2) Following these thoughts, social theories have a crucial influence on the definition of the tasks of planning and planning practice. Social developments on which the theories and perspectives are based, have a direct impact on attitudes, values, and understandings of roles of individuals and institutions socialised in this society. Consequences for planning practice and self-conceptions of planners can be derived from this. For instance, from a normative point of view, one could ask whether planners have the task of proclaiming de-growth and criticising prevailing paradigms.

Planning tasks as part of the challenges facing society as a whole Social theories relate to a broad understanding of planning, more comparable to the task and role of planning in society. Societal challenges also have an impact on planning, as planning is significantly involved in the implementation and preparation of decisions about space. This rather indirect interdependence is expressed, for instance, in the topic of digitisation. In itself, this project is not a dedicated planning task that has a direct spatial reference, but the provision of the necessary infrastructure (data, networks, etc.) and the methods for implementation imply a variety of planning tasks. In a way, the framework for the development is set here, which results from the content, the digitisation. This also illustrates the close connection between frame and content. While the framework represents the vehicle for the description and design of the planning process, it only exists depending on the dimension of content and the respective design. At the level of social theories, this fundamentally close interlocking seems rather diffuse, since these theories develop radiance and modes of action far beyond planning aspects. However, the two forms of theories go hand in hand, since both approaches, considering the framework without content, as well as considering the content without the framework, do not permit holistic statements. Social theories develop particularly clear frameworks that result from their theoretical content. Correspondingly, in the case of social theories it is almost impossible to talk about certain contents without also talking about the framework in which the implementation of the respective contents should be possible.

Holistic Planning Theories

The holistic description of what planning is Holistic planning theories try to describe planning as a holistic practice and to grasp it as an overall construct. These theories understand spatial planning as an agglomeration of actions, but without giving too much normative orientation. Rather, they are largely based on descriptive knowledge. This builds on the idea that systems and their properties should be viewed as a whole and not just as a composition of their parts. Planning and its practices are analysed in subcategories, but planning as an institution, practice and action cannot be fully understood from the interaction of its individual parts. This requires a superstructure, namely the strategy in strategic planning or the culture in planning culture. Holistic planning theories make an important contribution to planning sciences in many ways:

(1) Holistic planning theories are neither utopian in character nor directly action-oriented, but refer to descriptive-analytical forms of knowledge about spatial planning is. Some of the holistic planning theories also serve as a means of criticising existing planning practices or planning approaches. For instance, planning culture analyses planning systems and its cultural embeddedness in various European countries, showing the structural and institutional differences in the constitution of planning. Here, the impact of cultural practices can be understood and explained in conjunction with planning and its interpretation and understanding of roles.
(2) The strong foundation on descriptive knowledge and the understanding of what is understood as planning puts the empirical anchor of these theories in its very centre. This applies in particular to planning research, which aims to capture planning practice through the narrowest possible, empirically based evidence. Locally specific features and their characteristics play an important role in filling the respective planning theory. In this way, local practices and cultures make up the planning culture. Communicative planning thrives on the specific constellations of actors on site.
(3) It is striking that these holistic planning perspectives can be identified primarily in retrospect. By analysing past planning practices, actors, actions and processes can be understood, but it is rarely used to make predictions for planning and future implementations. In this way, communicative planning can be analysed after or within the planning processes, or planning-cultural configurations can be identified, but a direct future *ought* is not directly depicted in these planning theories.

Planning tasks as part of the holistic planning framework The claim of understanding planning as such in holistic theories results from the very close connection of framework and content. For instance, planning culture as theoretical approach uses cultural theory as framework (or lens) through which planning practices become tangible. The content is then those aspects that make up this culture, i.e., plans, the type of cooperation, common language or learning processes, etc. Accordingly, in holistic planning theories the interaction of framework and content becomes particularly important, due to their inherent holistic claim. Figure 6.2 shows that this meso level of planning theories is based on social theories, however, they usually hardly contain concrete recommendations for action due to its holistic view.

Theories of Action

Planning as organisation of actions and processes Planning theories that provide concrete instructions for action are described in the topology as action theories. They describe the planning procedure in concrete planning steps, phases and process flows. Planning theories, which can be read more or less concretely as instructions for action, are essentially based on regulative knowledge for the management and control of planning processes. Based on the rationalist approach of synoptic planning, some of the approaches were enhanced as a criticism of it, so to speak as a *reality check* of rationalist planning. This level of planning theories show how the future can be planned:

(1) Action theories guide actions more or less directly and are therefore geared towards practices. It is about how planning processes can be carried out and in most cases implies a narrow understanding of planning how the public sector plans. This becomes clear in the synoptic planning model or in incremental planning theories such as *muddling through.*
(2) The process orientation through process cycles or linear process designs is characteristic. Whether this is presented in illustrations of linear process steps or in circular, less linear processes with learning loops, is only important in the internal distinction. Due to the degree of concretisation, these theories of action have proven to be very useful for planning practice and in the teaching of planning sciences.

Planning tasks as recipes how to act Action theories translate theoretical frameworks into concrete instructions. The theoretical framing is not defined directly with it, one can rather imagine that the underlying social theory and the holistic understanding of planning embed the action approach. However, the social theories as well as holistic theories these action theories build upon are not deterministically specifying the course of action: Incremental planning, i.e., planning step by step, is based on the holistic idea that planning works as a political process and decision-making. Then planning as a political process provides the basic understanding, and incremental planning translates this into specific instructions for action. Another example is the concept of the 15-min city, which aims to make everyday routes accessible by foot. The calculation of accessibility is embedded in a rather rationalist theoretical framework, for which instruments such as distance calculations etc. are selected and applied. Action theories thus represent both framework and content, whereby both are highly interconnected with their embedding in social theories and holistic theories.

6.4 The Application of the Topology–Examples, Potentials and Limits

The Topology of Planning Theories in its application The Topology of Planning Theories is made up of The location in the field of planning knowledge and the degree of action-orientation. The location in the field of planning knowledge raises the question of whether the respective theory is based on a perspective closer to realism or one closer to social constructivism. In addition, the type of knowledge plays a fundamental role in the localisation. Is the focus on knowledge about the status quo or future-oriented knowledge more important for the theory? On the other hand, the theory's degree of action-orientation matters. Is it a planning theory that derives its basic assumptions from social theories and thus attempts to conceptualise central features of society? Alternatively, is it more of a planning theory that attempts to capture planning holistically as an overall construct and to present it accordingly? Or, does the theory postulate concrete instructions how or what to plan?

Significance of application A systematisation such as the Topology of Planning Theories only becomes understandable and clear in its entire scope when it is applied to different planning approaches and these become visible in the topology. The theories can be located in the field of planning knowledge, clear differences become apparent and similarities can be identified. It is also important to recognise the respective range and inherent knowledge orientation of the theories and what one wants to say with them in research or in practice. In the following, we have presented ten planning theories of different orientations as examples based on the topology in order to clarify our systematisation. In Fig. 6.3, they are located in the theoretical field (by the preceding numbers).

A Selection of Planning Theories Within the Topology

1. **Communicative planning** The approach of communicative planning aims at a strong involvement of citizens in planning processes. It assumes that influential information is that which has been socially constructed by the community that uses it. Planners appear here as participants in terms of their expert knowledge, but also as managers of different realities and truths within the planning process. Following this initial setting, the communicative planning theory takes a ***social-constructivist perspective*** on planning, to be located on the right side in

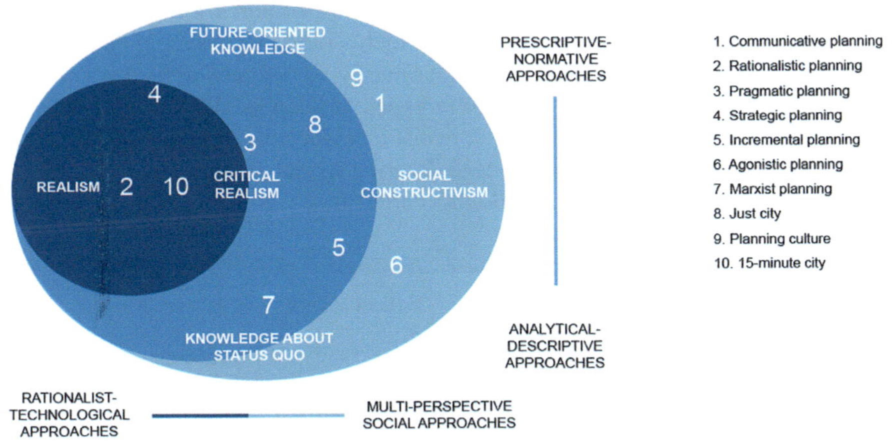

Fig. 6.3 Application of the Topology of Planning Theories. (own illustration)

the field of knowledge. At the same time, with regard to the types of knowledge, it accesses both future-oriented knowledge and knowledge about the status quo. Both play an important role in participatory processes with citizens. However, there is a stronger focus on a ***prescriptive-normative character,*** since future transformation and development is often the focus in debates with citizens and thus value questions about the desired future are addressed. Accordingly, communicative planning as a planning theory can be found in the upper right part of the theoretical field. The action-orientation of communicative planning can be described as a ***holistic theory*** and is fundamentally based on Habermas' discourse theory.

	Localisation in the field	Action-orientation	(Potential) embedding
Communicative planning	Social constructivist perspective Strong focus on future knowledge	Holistic theory	Action theory: e.g. Expert participation Social theory: Discourse theory

2. **Rationalist planning** The approach of rationalist planning, on the other hand, is located elsewhere in the field of planning knowledge. According to this theory, planning satisfies rational, factual, objective and scientific criteria. A core concern is the liberation of people from subjective and emotional structures as well as the help to control sovereignty over their environment by means of comprehensible and plausible decisions. Rationalist planning is characterised by a belief in the comprehensive planning of social, economic and spatial development. In addition, it is assumed that a rational order can be created and that solutions to the problems at hand exist. Accordingly, planning makes an important contribution to perfecting the world. The belief in objective knowledge, to which planners have access, forms the basis of rational decision-making processes (in

the sense of ***realism***) and describes both ***future-oriented knowledge*** and the status quo, but in particular the way in which a future state should be achieved. Accordingly, rationalist planning can primarily be assigned to the prescriptive-normative approach with regard to its types of knowledge. At the same time, as the description shows, it takes a realistic perspective on planning. Since the rationalist planning approach tries to describe spatial planning in its entirety, it can be assigned to the ***holistic planning theories***.

	Localisation in the field	Action-orientation	(Potential) embedding
Rationalist planning	Rationalist perspective Focus on future knowledge	Holistic theory	Action theory: e.g. synoptical planning model Social theory: realism/ rationalism

3. **Pragmatic planning** The pragmatic planning approach goes back to philosophy of pragmatism. This designates a current in which truth and theoretical knowledge arise from everyday actions and the importance of philosophical questions is measured by the practical, everyday consequences. After the failure of most theoretical approaches, pragmatic planning was initially understood from the 1980s as an "anti-theoretical" getting-things-done approach. In this approach, planning should apply the liberal virtues of reason, prudence and tolerance with regard to practical judgments and evaluations. Planners help to compare and select options in an intensive exchange with process participants. Plans serve as an imagined future and are guides for the practical evaluation of the respective projects. With regard to the types of knowledge, the pragmatic planning approach can be described as both ***descriptive-analytical and prescriptive-normative***. On the one hand, concrete instructions for action are developed, on the other hand, plans serve as long-term guidelines when evaluating actions. In doing so, pragmatic planning takes both a ***realistic and a social-constructivist perspective on planning***. In principle, realism is followed with regard to the existence of objects outside of human consciousness, but at the same time, the conditionality of the terms is recognised by common conventions. The approach of pragmatic planning can be assigned to the ***theories of action***, since the concrete procedure is primarily dealt with and information is given on how planners should behave in practice. However, there is a close connection to social theories on which the direct instructions for action are based - in this case on the philosophical current of pragmatism.

	Localisation in the field	Action-orientation	(potential) Embedding
Pragmatic planning	Social constructivist and realistic perspective, both present and future knowledge	Action Theory	Holistic theory: Planning as a political process Social theory: Pragmatism

4. **Strategic planning** From the late 1990s, strategic planning was very popular in practice. It should not only aim at formal documents, but also provide a framework for future decisions. The strategic planning compromises a clear goal orientation (in the form of models), shows a close connection between vision and action, makes a general attempt to organise the city/region as a learning system and is intended to provide orientation for local actors and activate and motivate key actors. Principles of business management were primarily adopted and linked to an incrementalism perspective. A distinction can be made between the linear and the adaptive strategy model. The former assumes a conscious decision-making process by rational actors, who plan and implement a strategy from top down based on comprehensive analyses. The latter shows that such linear strategy processes often fail. Strategies are much more likely to emerge from routine actions and spontaneous decisions. The aim is to recognise the strategies that develop in the process. With regard to its type of knowledge, strategic planning can be assigned more to ***future-oriented knowledge.*** There are certainly aspects that also allow a classification with regard to the knowledge about the status quo, however, the theoretical approach aims primarily at a medium to long-term goal and a strategy for achieving the respective goal is developed. At the same time, a ***realistic perspective*** on planning is taken, since the fixed structure of reality serves as the basis for developing strategies at all and for being able to use them successfully in a targeted manner to achieve goals. Strategic planning is a ***holistic theory,*** since the attempt is made to capture planning holistically.

	Localisation in the field	Action-orientation	(Potential) embedding
Strategic planning	Realism Perspective Strong focus on future knowledge	Holistic theory	Social theory: e.g management theory

5. **Incremental planning** The incremental planning approach focuses on planning by small, manageable and easily revisable steps. Planning is not understood as comprehensively rational, but as concrete, practical-political. Planning is constantly being reassessed and adjusted, and decisions are understood as negotiation processes. Instead of holistic planning, consensus-oriented realisations are carried out through projects in individual steps. This small-scale approach is also referred to as "muddling through". With regard to the type of knowledge, incremental planning is more likely to be assigned to ***status quo knowledge.*** This is mainly due to the practical-political and concrete orientation of the approach. Incremental planning takes on a more ***social-constructivist perspective*** on planning, since the respective individual practical-political dimensions are in the foreground when making decisions and implementing plans. Incremental planning can be described as ***action theory,*** since the concrete behaviour of planners is addressed here.

	Localisation in the field	Action-orientation	(Potential) embedding
Incremental planning	Social constructivist perspective Strong focus on knowledge of the status quo	Action Theory	Holistic theory: Planning as a political process, Social theory: Democracy theory

6. **Agonistic planning** The approach of agonistic planning developed primarily from criticism of communicative planning and criticism of the basic assumptions of discourse theory. Thus, the assumption of an ideal speech situation in a practical field is questioned and the inability to do justice to pluralism in society is used as an argument against reaching consensus. The variety of voices and the possibility of influencing the discourse prevent it. Representatives like Chantal Mouffe take a critical view of deliberative democracy anyway, since an antagonistic contradiction is at the heart of every debate and cannot be resolved by a discourse free of domination and an ideal speaking situation. Rather, the aim is to convert the antagonism into an agonism. This is a non-violent conflict in which dissent is a legitimate outcome, or consensus is always "conflicted consensus." With regard to its type of knowledge, the agonistic planning theory can be assigned primarily to ***knowledge about the status quo.*** Above all, existing structures are criticised here and fewer prospects and perspectives on the future are given. A ***social-constructivist perspective*** on planning is taken, which is reflected in the recognition of an individual and pluralised society with diverse subjective perceptions of the world. The approach can be described as a ***holistic theory,*** since planning is considered holistically.

	Localisation in the field	Action-orientation	(Potential) embedding
Agonistic planning	Social constructivist perspective Strong focus on future knowledge	Holistic theory	Action Theory: Agonism

7. **Marxist planning** Marxist planning is based on the socio-political flow of Marxism and above all criticises the assumption of rationally acting planning. Rather, planning is understood as an actor who conceals the underlying market logic by simulating rational decisions and processes. Moreover, planning is quite ineffective and has little impact on land prices, its allocation and inequalities. All this happens under the cloak of value freedom, which according to Marxist planning does not exist. In general, planning is viewed as market-oriented, where planning and the market work together and are not necessarily in conflict. In the existing sense, planning is therefore more of an instrument to support and protect the existing conditions and not an instrument to bring about changes. With regard to the type of knowledge, Marxist planning can be assigned to ***knowledge about the status quo,*** since the focus here is primarily on criticism of the status quo and less on normative assumptions about conditions to be achieved

in the future. In addition, following Marxism, Marxist planning adopts both ***a realistic perspective*** (regarding the existence of an objective reality independent of human reason) ***and a social-constructivist perspective*** (regarding a conclusion that does not necessarily follow an objectively *right* conclusion, planning is rather influenced by a human ideological self-deception). Marxist planning can be described as a ***holistic theory,*** due to the very close relationship to Marxism, but also as a ***social theory.***

	Localisation in the field	Action-orientation	(Potential) embedding
Marxist planning	Realistic as well as social constructivist perspective, strong focus on criticism of the status quo	Holistic theory, Social theory	Social theory: Marxism

8. **Just City** The Just City theory goes back to Susan Fainstein. Contrary to the perspective of seeing the city as a machine of (economic) growth, the Just City focuses on values such as justice, diversity and democracy. Accordingly, social justice is at the core of Just City and urban development projects should be evaluated from this point of view in particular. Cultural recognition and ethical demands of different social and ethnic groups are also an important pillar of the theory. With regard to the type of knowledge, the Just City approach can primarily be assigned to ***future-oriented knowledge.*** In particular, the normative stipulations in theory allow this classification, since future improvement and the development of alternative proposals for urban design play an important role here. At the same time, the approach takes a ***social-constructivist perspective*** on planning, since the conditionality of perception is considered an important theoretical component when considering different positions. This is a ***holistic theory*** that links social-theoretical questions (such as those about justice or what is meant by justice) with those relating to action (how should cities be redesigned so that they become just, etc.).

	Localisation in the field	Action-orientation	(Potential) embedding
Just City	Social constructivist perspective Strong focus on future knowledge	Holistic theory	Action Theory: Advocate planning Social theory: Justice discourse

9. **Planning culture** Approaches to planning culture attempt to draw attention to previously neglected aspects of spatial planning. For instance, on cultural elements of planning (such as values, norms, orientations and traditions), so that these are considered in depth or be analysed more empirically and a broader perspective can be taken that goes beyond public planning activities. In this way, spatial planning can also be analysed in a culturally comparative way. In addition, the planning culture approach differentiates itself from *Grand Theories* and takes a long-term perspective on forms of governance. The planning

culture approach is primarily concerned with ***knowledge about the status quo***, since existing knowledge is collected and compared here. A ***social-constructivist perspective*** on planning is taken, which attempts to represent and recognise the different cultural imprints. Due to the approach of gaining a holistic perspective on planning from different cultural backgrounds, the planning culture approach can be described as a ***holistic theory.***

	Localisation in the field	Action-orientation	(Potential) embedding
Planning culture	Social constructivist perspective Strong focus on status quo	Holistic theory	Action Theory: Good planning culture Social theory: Culture theory

10. **15-min city** The concept of the 15-min city envisages that all everyday routes that target basic needs should be accessible within 15 min by foot. Instead of the functional zoning of living, working and social areas, the functions should exist side by side and form a dense network. This is also achieved by a stronger mix of the buildings themselves - for example through the juxtaposition of apartments and service facilities. The mix is intended to help reduce inequalities by creating a polycentric urban structure that breaks with the monocentric orientation. In terms of types of knowledge, the 15-min city concept uses both ***status quo knowledge*** (since knowledge about existing inequalities and accessibility is needed) ***and future-oriented knowledge*** (since accessibility needs to be worked towards). A ***realistic perspective*** on planning is taken, since it is assumed that planning with the help of instruments and interventions can contribute to achieving this goal. The 15-min city can be described as an ***action theory,*** since a concrete goal is specified here, which is to be realised through the application of methods and instruments through planning action.

	Localisation in the field	Action-orientation	(Potential) embedding
15 min city	Realistic perspective both knowledge of the status quo and future knowledge	Action Theory	Holistic theory: rationalist planning Social theory: rationalism

Potentials and Limits of the Topology of Planning Theories

What the Topology of Planning Theories can provide As becomes clear from the application of topology, this approach to systematisation differs significantly from previous ones—with the aim of emphasising the knowledge inherent in planning theories. The location in a planning-theoretical field and the degree of action-orientation correspond to the claim for simplification through theory, but also the presentation of complex interdependencies and interactions. In general, the links and relationships between the theories are more important in this topology than the separating elements. The links are recognisable both in the field of planning knowledge and through the references between the different levels of action-orientation in planning theories.

Locating theories in a field of knowledge The subdivisions made in the field of planning knowledge should not be perceived as too rigid, but rather a relative classification into rationalist and/or social-constructivist perspectives. Other approaches, such as that of critical realism, are not explicitly stated here as a position, but can be located in the middle as a mixed form from an ontological point of view. Even if the normative-ethical dimension of planning theories is not directly represented in a separate area, it still plays an important role in the forms of knowledge, especially in future-oriented knowledge, which is always normative. In this regard, the distinction between the different levels of action-orientation seems to be very exciting, as it represents an in-depth examination of the basic assumptions of the respective planning theory. This becomes particularly obvious in the social theories, which present the perspective and framework of the theory, but also in the holistic theories, which clarify the underlying understanding of planning (and its peculiarities). Action theories, on the other hand, show how these frameworks (can) be organized in a very concrete way, since they already formulate recommendations for action, which can be classified in a smooth transition to models, concepts and models. Despite the strong action-orientation, they are seen as theoretical approaches, they are also called theories (such as Christaller's central place theory)–so a differentiation seems helpful here. But not only the classification of individual theories is interesting, also the considerations of which social theories justify the theories or which holistic theory is the starting point of the theory used is enormously helpful for research and practice. This corresponds to the idea of building this pyramid from a theory: Which theory justifies the respective planning theory? Which theory can be set up as an action theory? It is important here that this does not happen in a deterministic manner. For example, incremental or pragmatic theory can be applied to holistic planning as a political process. The specific pyramid of theoretical direction needs to be reassembled for each approach to research or planning practices. The examples given here are also only to be interpreted in this way. It is doubtful whether every approach requires a direct action theory. It has already become clear that the focus of a theory on the status quo (in the field) usually means that no direct theory of action can be determined - this also makes sense, because without direct future orientation, based solely on analytical knowledge, no action can be derived.

Potentials and limits of topology The Topology of Planning Theories certainly does not solve all the challenges of existing systematisation approaches or unite them all into a common approach. However, it provides new perspectives on the links between planning theories and the different perspectives inherent in them. To do this, it uncovers the ontological and epistemological backgrounds of planning theories and connects them with the respective action-orientation. In this way, the dense interrelations of theoretical backgrounds with the practical effects of the theories becomes transparent. The value of this topology and at the same time its different perspective becomes particularly obvious in a change of perspective, the reflection on planning theories beyond the dichotomy of substantive theories and procedural theories. A closer look into relevant planning journals where theoretical approaches are discussed reveals that substantive theories are rarely addressed. It seems obvious that these substantial theories have hardly found their place in the theoretical discourse–at least in the discourses that are also called planning theory. The topology of the planning theories attempts, in a first approach, to integrate them as subject of theoretical considerations again. This does not mean to concentrate only on substantial theories now, but to acknowledge the simultaneity of substantial and procedural components within planning theories, providing the procedural as well as content-specific issues. For instance, the theory of a Just City as a holistic theory provides a rather social-constructivist perspective and demands to integrate different views and opinions especially of those rarely heard (procedural dimension). Content-wise, the theory could then go even further to illustrate how the term of justice is to realize in terms of access to infrastructure, services, etc.. These linkages of planning processes and content-related aspects are not fully worked out with this book; on the contrary, we see this book as an invitation to join this experiment and to strive for a stronger foundation of all theories and forms of knowledge in spatial planning. This advancement of planning theories is not done for its own sake, but to make planning knowledge explicit, open for discussion and to develop it further. In first applications within a seminar with phd-students, the topology and especially the location in the field of knowledge helped crucially to understand one's own approaches, to reflect one's own understanding of realities, of planning, and of one's own research. If a person chooses, for example, actor-network theories with a strong analytical knowledge orientation for the research of spatial planning, the research does not end directly in recommendations for action (as future-oriented knowledge), at least not without further steps. This requires another theoretical approach, in particular with a focus on action theory, so that one does not fall into the trap of the naturalistic fallacy. How far the Topology of Planning Theories may help to differentiate between theoretical approaches has to be proven in practice as well. In any case, it represents an enhancement and in-depth reflection beyond the types of knowledge that we now classify as planning theories. The Topology of Planning Theories calls into question three current paradigms of planning sciences: The dichotomous distinction between process and content, the gap between theory and practice, and the insecure handling of the strong action-orientation of the discipline.

References

Behrend L, Levin-Keitel M (2020) Planning as scientific discipline? Digging deep toward the bottom line of the debate. Plan Theory 19(3):306–323. https://doi.org/10.1177/1473095219897283

Faludi A (1973) Planning theory. Repr. Pergamon Press (The urban and regional planning series, 7), Oxford

Hoyningen-Huene P (2013) Systematicity—the nature of science. Oxford University Press, New York

Poser H (2012) Wissenschaftstheorie, 2. überarbeitete und erweiterte Aufl. Reclam, Stuttgart

What Can Planning Achieve? 7

Abstract

The conclusions that can be drawn for planning as a discipline and professional field from a reorganisation of planning knowledge give rise to hope. The presentation of the diversity of planning knowledge beyond procedural theories and their genesis can serve as a basis for urgently needed spatial and planning expertise for the challenges of the twenty-first century.

Keywords

Planning knowledge • Paradigms of planning sciences • Future challenges

Topology as a result of debate Spatial planning in science and practice is incredibly diverse, as diverse as the theoretical foundations with which we look at it. At the beginning of this book, we started by outlining the main goals of the topology and why the idea appeared in our thoughts: taking a perspective taking up a more precise definition of planning knowledge, making basic assumptions and understandings of planning and planning theories explicit, and underlining their characteristic reference to action. This is how the Topology of Planning Theories came about, as a systematisation that reflects the fundamental importance of planning theories, and as planning knowledge revealing the foundations of spatial planning. In other words, the topology is only one part of the topics discussed in this book, additionally, we examine planning foundations and its explanations from a perspective of philosophy of science–and thus question some of the main paradigms in planning theory and its debates.

M. Levin-Keitel and L. Behrend, *The Topology of Planning Theories*, The Urban Book Series, https://doi.org/10.1007/978-3-031-37857-7_7

7.1 Specification of Planning Knowledge on Many Levels

Different understandings of planning One approach to specifying planning knowledge is to deal with spatial planning and its different understandings, i.e. the epistemic object or subject of this knowledge. The need to distinguish between different understandings of planning is fundamental to any further understanding of planning theories. The variety of planning understandings reflects planning as such and its diverse practices. This diversity enriches the field of spatial planning, but becomes a challenge in the communication and concrete description of what we imagine planning to be, which planning practice is assumed, which practices, activities and actions are to be understood by planning and to what extent these practices condition, reflect, and specify our research approaches. Accordingly, definitions are always only an attempt at approximation and do not have a universally claim to validity. Depending on a narrow or broad understanding of planning, the answer to the question of what is planned or how it is planned differs. Spatial planning in a narrower sense focuses strongly on the question of the ability of the public sector to manage and control spatial processes. In a broad sense, spatial planning aims at participation in a jointly designed spatial development. As we have been able to show, this specification of spatial planning also contains an ontological distinction, namely the question of classification in a rationalist or social-constructivist perspective. In a narrow understanding of planning, the question of *what* spatial planning plans is a question of controlling and regulating public space and its uses. In a broad understanding, the question of *what* is planned can be mapped much more broadly as the influence of spatially effective actions in social space and thus society. Here, too, there are fundamental differences, depending on the understanding of planning. In retrospect, spatial planning in planning theories and planning sciences up to the postmodern times was primarily understood as a state task in a narrow understanding of planning.

Reveal basic theoretical assumptions Scientific knowledge is specified in form of theories, in our case planning theories and their description of a theoretical field. With the Topology of Planning Theories, we create a tighter interlocking of planning theory with regard to its theoretical assumptions and its reference to action than has been the case up to now. In previous approaches to the systematisation of planning theories, dichotomies are usually opened up, such as in the divide of substantive and procedural theories. As fruitful as a simplification and delimitation can be on a theoretical level, it reflects the connections in practice insufficiently and does not do justice to the interdependencies in theory. As has already been mentioned, not only do the different understandings of planning reveal an epistemological position (e.g. with the question of how many realities are recognised), the dimensions of knowledge contained in planning theories also play a fundamental role for other basic theoretical assumptions. Based on a detailed presentation of both the foundations in philosophy of science (realism and social constructivism, etc.) as well as the different types of knowledge in planning (future and present, ethical-normative orientation, etc.), we developed an instrument to reflect upon the landscape of planning theories. By locating planning theories in the field of knowledge it is possible to reveal their individual basic assumptions and thus inherent potentials and limits, especially in their relationships to each other. This supports the prominent role of definitions and clarification of terms in the scientific debates. In particular, the basic assumption of critical realism, which represents the intellectual basis of this elaboration, supports this view on the role of definitions: we assume that the world is recognisable in principle, that we basically assume objects that exist independently of observation and about which we can produce knowledge (ontological perspective of realism). However, as humans we do not have direct access to these objects, but perceive reality through our senses and our mental apparatus, i.e. a subjective processing of it (epistemological perspective of social-constructivism). As a result, every consideration of planning practice is cognitively distorted and already filtered through theoretical lenses, sorted and not insignificantly shaped–in other words: practice and theory are closely and inevitably linked.

The characteristic action-orientation The characteristic action-orientation of planning sciences and in various degrees of planning theories is one of the key dimensions, which has crucial importance and is subject to various interpretations, especially in scientific debates. The action-orientation and the strong reference of the planning sciences to the professional field of planning are based on some rarely discussed basic assumptions. Main question is how planning sciences and their theories influence of the profession or in how far the profession should be the addressee of planning theories. In planning publications, it differs from the role of a simple description of planning as the practice of knowledge, the advancement of theory serving to get closer to practice, or the postulating the (undesired) gap between theory and practice. However, we adopt a different perspective towards the relationship between theory and practice, a basic attitude that assumes a unity between theory and practice. If one describes theories as a lens or glasses with which one approaches reality out there, then theory and practice are two sides of the same coin, irrevocably connected and they depend on each other, there is no practice without any theoretical glasses. Even if you are not necessarily always aware of the theoretical basic assumptions, they still influence what you perceive as reality. This means that theories and practices are a direct part of an overall view, and the question of action-orientation of planning theories is completely different: Which theory has which action-orientation and what is this (implicitly) based on? Could the knowledge contained in the theories then be classified? As a logical consequence, action-orientation is also an important component of our Topology of Planning Theories: The differentiation of levels of action-orientation into the social macro-level, a (neo-)institutional meso-level and an action-oriented micro-level. Of course this differentiation manifests itself both through the distinction between narrow and broad understandings of planning (i.e. who is seen as a planner), as well as the differentiation of different disciplinary backgrounds (i.e. the theory has roots in social science, political science or economics, for example). The disciplinary roots of the respective planning theory decisively define the respective level of action, since they relate, for example, to a society and thus planning as a social task, to political processes and thus planning from a political perspective, to economic aspects and thus planning as part of the economic system and foundations for developing planning practice. The differentiation into narrow or broad planning understandings determines the action-orientation, since it is not insignificant who is supposed to carry out these actions.

7.2 Questioning the Paradigms of Planning Sciences

Paradigms and the foundations of the topology The Topology of Planning Theories challenges fundamental assumptions in planning sciences which, in our opinion, go hand in hand with an intensive study of epistemological foundations. Firstly, there is the rigorous separation of theory and practice, which is reflected in so many aspects of the theoretical discussion in planning sciences: from the gap that needs to be measured, to the attempt to develop a theory that corresponds more closely to practice or even the denial of planning theories in general, as they have no practical use. Secondly, we have–also in our own thoughts about planning theories and how we teach them - found the dominance of the division of theories into process and content, which forbids certain avenues of thought and which we also found difficult to overcome in relation to the topology. It is not only about the separation, which already sets different directions of thought and makes others more difficult, but in particular about the dominance of the discussion about procedural theories and the rather underexposed knowledge of substantial theories.

The separation of theory and practice In the planning literature and in planning-theoretical discourses, the gap between theory and practice is constantly mentioned and its existence is always assumed to be a given foundation. We have explained in this book why, according to our argument, this gap does not exist. A practice that is independent of theory is obsolete, since every action always uses underlying theoretical assumptions. These do not have to be obvious, not even to the acting person him- or herself, but they are always present. Even basic assumptions such as the one that phenomena exist independently of the human mind and that planning as such exists (as planning is not dependent on human presence), are inherently theoretical in nature. However, theoretical assumptions are the core of any perception of our surroundings. For example, when managing spaces, there are always questions of justice that appear, in terms of distribution and use. Even if it is assumed that decisions and considerations can be understood rationally and objectively, there are always normative settings about the target state, which is considered desirable. In addition, planning always aims at future conditions and thereby conveys assumptions about the ability to plan. At the same time, the selection of instruments and methods is both driven by values and shaped by fundamental assumptions about their effectiveness.

The separation of process and content Secondly, we did not follow up on the strict divide into substantive and procedural planning theories. The Topology of Planning

Theories differentiates according to the respective research object, i.e. the epistemological object (or subject) of spatial planning in terms of content and process, as the separation of these is only an analytical one. They are in constant exchange and shape each other. The presented Topology of Planning Theories introduces a systematisation that reduces the theories in planning to their basic components. With the attempt to break down the different epistemological aspects of planning theories and thus to describe theoretical fields in planning sciences, the modes of action and validity claims of theories are made explicit and existing dichotomous representations are at least weakened. The debate about a gap between theory and practice can also be viewed from the perspective of the separation of procedural and substantive. While in theory there is a focus on the process of planning, practitioners primarily need applicable instruments and proposed solutions to everyday problems and planning content. The gap is therefore less a gap than a communicative misunderstanding, since academics like to talk *about planning* and planners as part of it, while practitioners talk more substantively about *planning in its concrete form* and seek direct reference to action. The topology of planning refers to this misunderstanding by linking theory and action/practice more closely. In addition to these advantages, however, there is also a challenge that cannot be easily overcome by the Topology of Planning Theories. The demand to overcome dichotomies poses difficulties for the topology. Although this can be used to show the multitude of links within and between planning theories, the presentation of the planning field of knowledge also falls into a dichotomy-like subdivision. However, this divide is less rigid than, for instance, a distinction of substantial and procedural planning theories in general. Rather, the links between the different theories is shown. Thus, the Topology of Planning Theory contributes significantly to the dismantling of this divide and locates planning theories in relation to each other, a novel way to understand the landscape of theories in previous systematisation approaches.

7.3 Planning Theories for Future Challenges

Planning theories and their future significants In the debates on planning theories, several paradigm shifts have repeatedly taken place over time, from rationalist planning to communicative planning and further. These shifts are fundamental, due to changes in the concepts of knowledge and the underlying normative assumptions. Most obvious in the transition from rationalist planning–planners know what is right for society and which processes lead to an improvement in living conditions–to a communicative understanding of planning–planners are only part of the decision-making and need to collaborate with other actors, world-views and problem conceptualisations. This shift has been discussed in the literature as communicative turn and is the basis of many planning debates. It remains controversial to this day whether this paradigm shift was actually completed or whether it is still in a pre-paradigmatic stage. With the progress of the debate and the accompanying criticism of the basic assumptions and practical feasibility of communicative planning theory, the approaches have meanwhile multiplied. In particular, agonistic and pragmatic perspectives are more and more discussed. Even if, for instance, approaches of pragmatism are increasingly popular, we see the persistence in the current paradigmatic basic assumptions as problematic. There is a multitude of paradigms in spatial planning that shape the picture of theories and practice. Just as lenses influence the sharpness of glasses, they shape a specific image of the object of planning. We have now tried to shake up some of these basic assumptions and thus enable new perspectives and points of view–precisely because planning knowledge must be essential for the great challenges of a sustainable way of life.

Planning and transdisciplinary research The Topology of Planning Theories represents a reorganisation of knowledge for planning in face of the challenges in the twenty-first century. Numerous challenges, be it the effects of climate change and adaptation and avoidance strategies or social polarisation and the question of appropriate decision-making, require a rethinking and a reorientation of planning approaches. Facing numerous complex problems, finding inter- and transdisciplinary solutions becomes increasingly important for a society in transition. Therefore, the integration of different forms and types of knowledge gains crucial attention, from a scientific view this requires an ontological turn–and a reconsideration whose knowledge is to be integrated at which point. As we have shown, this could directly refer to some of the shifts made in planning theory, as in these discussions assumptions about the knowledge basis of decisions have been at issue (rationalist planning = planner as carrier of knowledge; communicative planning= majority as carrier of knowledge). The Topology of Planning Theories makes this ontological differentiation visible, locating theories within the topology represents an important first step to point out where this knowledge of planning is rooted. It enables a systematic presentation of different approaches to planning theory, which often stand side by and are compared with one another, although, as becomes clear in the topology, they have different goals and practical orientations. In this respect, the topology of the planning theories is not to be seen as the solution to the problem, but as a preparatory part and a possibility for an epistemological positioning. It represents a tool to unravel and order the different threads of planning theory. Building on this, further theoretical considerations can then be made, which can contribute to the relevant understanding of spatial planning. This book does not offer answers and solutions in the general sense, but asks questions, enables other perspectives on previous planning theories and their knowledge content and restructures them. Planning knowledge as well as the strong action reference of the planning sciences are discussed, further formulated and unusual perspectives on known approaches are articulated.

Glossary of Planning Theories

The glossary of planning theories gives an initial overview of what each theory contains from an epistemological perspective and how it can be classified in the Topology of Planning Theories. Arranged according to their relation to action, we briefly present the selected theories and reveal their planning knowledge. In doing so, we do not go into social theories such as discourse ethics or Marxism, although these serve as basic theoretical framework, since due to their mostly missing reference to planning, they are mostly applied and transferred to planning in the form of holistic planning theories.

Holistic Planning Theories

Agonistic planning The approach of agonistic planning derives primarily from criticism of communicative planning and criticism of the basic assumptions of discourse theory. Thus, the assumption of an ideal speech situation in a practical field is questioned and the inability to do justice to pluralism in society is used as an argument against reaching consensus. Based on agonism and the assumption that an antagonistic contradiction is at the heart of every debate, a discourse free of domination and an ideal speech situation are doubted. Rather, the aim is to convert the antagonism into an agonism. This is a non-violent conflict in which dissent is a legitimate outcome, or consensus is always *conflicted consensus.* The theory adopts a social-constructivist perspective, strongly oriented towards present knowledge with weak future orientation. Due to the little action-guiding character, agonistic planning represents a holistic theory.

Further Reading

Pløger J (2004) Strife: Urban planning and agonism. Plann Theor 3:71–92
Roskamm N (2015) On the other side of „agonism ": „The enemy ", the „outside ", and the role of antagonism. Plann Theor 14:384–403.

M. Levin-Keitel and L. Behrend, *The Topology of Planning Theories*, The Urban Book Series, https://doi.org/10.1007/978-3-031-37857-7

Feminist/anti-racist planning Feminist planning questions gender categories and the associated attributions. The concept of gender planning also established itself as (gender) inclusive urban planning, in which a large number of perspectives on the city and planning are to be integrated. In this way, the different life plans should be met. Anti-racist planning primarily considers the structural injustices and exclusions that arise through planning measures and institutional actions. From a social constructivist perspective, these planning theories criticise the status quo and thus identify planning deficits in society. They reveal little action-guiding knowledge, which is why they are to be classified as holistic theories.

Further Reading

Sandercock L, Forsyth A (1992) A gender agenda: new directions for planning theory. J Am Plann Assoc 58(1):49–59.

Huning S, Mölders T, Zibell B (2019) Gender, space and development. An introduction to concepts and debates. In: Zibell B, Damyanovic D, Sturm U (ed) Gendered approaches to spatial development in Europe—perspectives, similarities and differences. Abingdon, New York, p. 1–23.

Governance Approach Governance is the generic term for all forms with which the collective and interdependent actions of actors and organisations can be coordinated. The focus here is on the institutions and the regulatory structures that prevail in them. In addition, there is a task of analytically separating control subject and control object. State, market and networks are understood as complementary forms of control and decision-making processes beyond the state institutional system are analysed accordingly. This corresponds to a social constructivist perspective in which more than one reality is assumed and integrated into the planning process. In comparison, for example, to rationalist approaches, structural, functional and instrumental aspects of governance, control and coordination are examined in addition to process-related ones.

Further Reading

Duit A. and Galaz V (2008) Governance and complexity—emerging issues for governance theory. *Governance*, 21(3), 311–335.

Mayntz R. (1998) New challenges to governance theory. Governance as social and political communication. Jean Monnet Chair Papers 50, The Robert Schuman Centre at the European University Institute.

Just City The Just City theory goes back to Susan Fainstein. Contrary to the perspective of seeing the city as a machine of (economic) growth, the Just City focuses on values such as justice, diversity and democracy. Therefore, urban space and the way in which the burdens and benefits of development are distributed spatially, and the decision-making processes that accompany this distribution, are a central concern of spatial justice. Accordingly, social justice is at the core of Just City and urban development projects should be evaluated from this point of view in particular. Indeed, spatial justice focuses on the spatial dimensions of the equitable distribution of the burdens and benefits of development (distributive justice) and on the particular way in which this distribution is managed and decided (procedural justice). Cultural recognition and ethical demands of different social and ethnic groups are also an important pillar of the theory.

Further Reading

Uitermark J, Nicholls W (2017) Planning for social justice: strategies, dilemmas, trade-offs. Plann Theor 16(1):32–50.

Fainstein SS (2011) The just city. Cornell University Press.

Communicative planning The idea of communicative planning is that planning action is guided by norms and that communication and cooperation between different actors lead to an understanding of planning issues—through discourse and the exchange of arguments. The communicative planning approach aims to involve citizens more closely in the planning processes. It assumes that influential information is that which has been socially constructed by the community that uses it. Planners appear here as participants in terms of their expert knowledge, but also as managers of the planning process. At its core, supporters of this approach assume that planning will become more democratic, more legitimate and more just by improving communication between the actors.

Further Reading

Innes JE (1995) Planning theory's emerging paradigm: communicative action and interactive practice. J Plann Educ Res 14(3):183–189.

Healey P (1992) Planning through debate: the communicative turn in planning theory. Town Plann Rev 63(2):143–162.

Marxist planning Marxist planning is based on the socio-political flow of Marxism and above all criticises the assumption of rationalist planning. Rather, planning is understood as an actor who conceals the underlying market logic by simulating rational decisions and processes. Moreover, planning is quite ineffective and has little impact on land prices, its allocation and inequalities. All this happens under the cloak of value freedom, which according to Marxist planning does not exist. In general, planning is viewed as market-oriented, where planning and the market work together and are not necessarily in conflict. In the existing sense, planning is therefore more of an instrument to support and protect the existing conditions and not an instrument to bring about changes. It incorporates elements of a social-constructivist and a rationalist perspective.

Further Reading

Harvey D (1985) The urbanisation of the capital. Johns Hopkins University Press, Baltimore.

Dear, M, Scott, A (1981) Urbanisation and urban planning in capitalist society. Methuen, London.

Planning culture Theoretical approaches of planning culture attempt to draw attention to previously neglected aspects of spatial planning. This includes understanding planning as a cultural act and taking greater account of cultural elements of planning (such as values, norms, orientations and traditions). The aim is to examine planning on the basis of its practices in an empirically founded manner and to take a broader perspective beyond planning under public law. In this way, spatial planning can also be analysed in a culturally comparative manner, at national levels as well as locally. The way in which the respective actors understand their roles and tasks, how they perceive problems, deal with them and apply certain rules, procedures and instruments–from a social constructivist perspective–are thus characteristics of a planning culture.

Further Reading

Bishwapriya S (Ed.) (2005) Comparative planning cultures. Routledge.

Othengrafen, F and Reimer, M (2013) The embeddedness of planning in cultural contexts: theoretical foundations for the analysis of dynamic planning cultures. Environment and Planning A, 45(6), 1269–1284.

Rationalist planning Rationalist planning is based on rational, factual, objective and scientific criteria. Deeply rooted in rationalism, the liberation of people from subjective and emotional structures as well as help to control their environment by means of comprehensible and plausible decisions is a core concern. Rationalist planning is characterised by a belief in the comprehensive planning of social, economic and spatial development. Accordingly, planning makes an important contribution to perfecting the world. The belief in objective knowledge, to which planners have access, forms the basis of rational decision-making processes (in the sense of realism) and describes future-oriented knowledge more than a status quo, in particular the way in which a future state is to be achieved.

Further Reading

Meyerson M, Banfield EC (1955) Politics, planning and the public interest—the case of public housing in Chicago. Free Press, London.

Faludi A (1986) Critical rationalism and planning methodology. Pion, London.

Strategic planning Strategic planning not only aims at formal documents and strategy papers, but also provides a framework for future decisions. It has a clear goal orientation (in the form of models), shows a close connection between vision and action, makes a general attempt to organise the city/region as a learning system and motivates and activates key actors. A distinction can be made between the linear and the adaptive strategy model. The former is based on a conscious decision-making process by rational actors, who build on comprehensive analyses and plan and implement a strategy top-down. The latter shows that such linear strategy processes often fail. Strategies are much more likely to emerge through routine actions and spontaneous decisions. The aim is to recognise the strategies that develop in the process.

Further Reading

Salet W, Faludi A (ed) (2000) The revival of strategic spatial planning. Elsevier, Amsterdam.

Mintzberg, H (1993) Strategic Planning. California Management Review, 36(1), 32–47.

Theories of Action

Advocacy Planning Advocacy planning stands for social and economic justice and sees planners as representatives of marginalised and excluded groups. Planners should ensure that the interests of these groups are integrated into the planning processes. Accordingly, in addition to their professional expertise, planners also take on this separate role as advocates for these groups or the environment. This theory adopts a more social-constructivist perspective that has a strong normative character, based on the analytical knowledge that certain groups are less heard in spatial planning. As an action theory, advocatory planning refers to a direct action orientation, namely advocating for precisely that group.

Further Reading

Davidoff P (1965) Advocacy and pluralism in planning. J Am Inst Planners 31:331–338

Athens Charter The Athens Charter, developed at a congress in 1933, represents a modernist urban planning program. The focus is primarily on the functional structure, while urban design only plays a subordinate role. Above all, the charter formulates an answer to the poor living conditions by ensuring light, air and sun through a high density of housing in high-rise buildings and the surrounding spacious open spaces. Through the demolition of slums and the redesign, the recreational function (by open spaces), the work function (separation of living and work, but in the shortest possible distance) and mobility (by motorised private transport and the corresponding infrastructure for this) should be ensured. Here, planning is largely done in the perspective of rationalism.

Further Reading

Gold, J R (1998) Creating the Charter of Athens: CIAM and the functional city, 1933–43. *The Town Planning Review*, 225–247.

Decentralised concentration The model of decentralised concentration represents a further development of Walter Christaller's system of central places. It seems to represent a unique German planning theory on how to develop a region. Unlike Christaller's focus on centres along transport hubs, this model focusses on the periphery from a regional perspective. In particular, a framework is set out for the development of the areas surrounding cities. Accordingly, the model of decentralised concentration is made for metropolitan regions and tries to shape them in the sense of a mixed and compact settlement development. While concentration is being sought at the small-scale level, decentralisation is to be established at the large-scale level. Accordingly, a large number of centres are emerging, which combine social and economic functions. At the same time, urban sprawl is avoided and controlled, all kinds of development primarily takes place in and around these centres.

Further Reading

Priebs A (2010) Dezentralisierte Konzentration. In Henckel D, von Kuczkowski K, Lau P, Pahl-Weber, E, Stellmacher F (ed) Planung—Bauen—Umwelt. Ein Handbuch. Wiesbaden, p. 109–112.

European City The model of the European City emerged in the 1990s from the linking of the historical quality of European cities and the sustainability discourse. At the same time, it forms the final replacement of the concepts that were developed from urban modernism. The *European City* represents a symbol for an urban ideal type, which is based on the historically developed European cities. The task of planning is to create the spatial conditions for energy and resource efficiency, especially with regard to climate change and adaptation. Accordingly, the material and energy cycles associated with spatial developments must be taken into account and efficient use of the existing infrastructure should be ensured.

Further Reading

Häussermann, H (2005) The end of the European City?. European Review 13(2), 237–249.

Garden City The model of the garden city goes back to the British Sir Ebenezer Howard. His work *Garden Cities of Tomorrow,* published in 1902, served as the basis for the realised garden city of Letchworth. He developed his garden city concept (1902) as a so-called "city-country-city" and opposed the urban growth as an alternative spatial development. The core of the concept is the strict separation of functions, which are implemented within the concentric rings of the city. The functions of supplying, working, living and relaxing are arranged in close proximity to one another. This should also reduce inner-city traffic. Commercial establishments are located on the outskirts and are connected by railway networks. In addition to the extensive mix of functions and the urban structure of the small house with garden, land reform with the orientation towards communal and cooperative land ownership played a central role in Howard's model.

Further Reading

Howard E (1965) Garden Cities of Tomorrow, Bd 23. With Press.

Ward, S (2005) *The garden city: Past, present and future.* Routledge.

Incremental planning The incremental planning approach focuses on controlling small, manageable and easily revisable steps. Planning is not understood as comprehensively rationalist, but as concrete, practical-political. Plans are constantly being reassessed and adjusted, and decisions are understood as negotiation processes. Instead of holistic planning, consensus-oriented realisations are carried out through projects in individual steps. This small-scale approach is also referred to as "muddling through". Incremental planning takes on a more social-constructivist perspective on planning, since the respective individual practical-political dimensions are in the foreground when making decisions and implementing plans.

Further Reading
Lindblom CE (1959) The science of muddling through. Public Adm Rev 19(2):79–88.

Pragmatic planning The pragmatic planning approach goes back to philosophical pragmatism. This designates a current in which truth and theoretical knowledge arise from everyday actions and the importance of philosophical questions is measured by the practical, everyday consequences. After the failure of most theoretical approaches, pragmatic planning was initially understood from the 1980s as an "anti-theoretical" getting-things-done approach. In this approach, planning should apply the liberal virtues of reason, prudence and tolerance with regard to practical judgments and evaluations. Planners help to compare and select options in an intensive exchange with process participants. Plans serve as an imagined future and are guides for the practical evaluation of the respective projects.

Further Reading
Forester J (1982) Planning in the face of power. J Am Plann Assoc 48(1):67–80
Flyvbjerg B (2002) Bringing power to planning research: one researcher's praxis story. J Plann Educ Res 21(4):353–366.

Synoptic planning model A planning process can be understood as synoptic, which has a clear starting point and a clear goal and includes the steps of planning and implementation on the way there. The synoptic planning model is linear, comprehensive, and sees planning as a self-contained process, from problem definition through information gathering to problem resolution. The omniscient and omnipotent role of the planner is at the centre of the implementation of planning. The synoptic planning corresponds to the model of the integrated and comprehensive development planning (or *comprehensive planning*). A fundamentally rationalist understanding of planning is characteristic of this approach. The synoptic planning model falls under the action theories, since there are direct instructions for action for implementation.

Further Reading
Hudson, B M, Galloway, T D, & Kaufman, J L (1979). Comparison of current planning theories: Counterparts and contradictions. *Journal of the American planning association*, *45*(4), 387–398.

Innes J (1996) Planning through consensus building: a new view of comprehensive planning ideal. J Am Plann Assoc 62(4):460–472.

Central Places Concept Going back to his Central Places Theory (1933), Walter Christaller developed the Central Places Concept, which is still an important spatial planning concept in Germany today. Centrality results here from the functional aspects and not from a geometry. In his analytical model, he developed a structure of central locations on different hierarchical levels. The central locations at higher hierarchical levels (e.g. larger cities) have features that the central locations at lower hierarchical levels lack (e.g. certain administrative and service functions). Functional aspects exist, for example, with regard to a supply, a traffic or an administrative function of cities and places for their surrounding areas. The supply function is the most important.

Further Reading

Mulligan, G F (1984) Agglomeration and central place theory: A review of the literature. *International Regional Science Review*, *9*(1), 1–42.

Eaton, B C, & Lipsey, R G (1982) An economic theory of central places. *The economic journal*, *92*(365), 56–72.

15-min city The concept of the 15-min city envisages that all everyday routes that target basic needs can be reached on foot within 15 min. Instead of the functional separation of living, working and social areas, the functions exist side by side and form a dense network. This is also achieved through a stronger mix of the buildings themselves–for example through the juxtaposition of apartments and service facilities. The mix helps to reduce inequalities by creating a polycentric urban structure that breaks with monocentric orientation. A realistic perspective on planning is taken, since it is assumed that planning with the help of instruments and interventions can contribute to achieving this goal.

Further Reading

Moreno, C, Allam, Z, Chabaud, D, Gall, C, & Pratlong, F (2021) Introducing the "15-Minute City": Sustainability, resilience and place identity in future post-pandemic cities. *Smart Cities*, 4(1), 93-111.

Pozoukidou G, Chatziyiannaki Z (2021) 15-min City: Decomposing the new urban planning utopia. Sustainability 13:928

Index

M. Levin-Keitel and L. Behrend, *The Topology of Planning Theories*, The Urban Book Series, https://doi.org/10.1007/978-3-031-37857-7

GPSR Compliance

The European Union's (EU) General Product Safety Regulation (GPSR) is a set of rules that requires consumer products to be safe and our obligations to ensure this.

If you have any concerns about our products, you can contact us on ProductSafety@springernature.com

In case Publisher is established outside the EU, the EU authorized representative is:

Springer Nature Customer Service Center GmbH
Europaplatz 3
69115 Heidelberg, Germany

Batch number: 10732090

Printed by Printforce, the Netherlands